T0137176

Advanced Sciences and Technologies for Security Applications

Indexed by SCOPUS

The series Advanced Sciences and Technologies for Security Applications comprises interdisciplinary research covering the theory, foundations and domain-specific topics pertaining to security. Publications within the series are peer-reviewed monographs and edited works in the areas of:

- biological and chemical threat recognition and detection (e.g., biosensors, aerosols, forensics)
- crisis and disaster management
- terrorism
- cyber security and secure information systems (e.g., encryption, optical and photonic systems)
- traditional and non-traditional security
- energy, food and resource security
- economic security and securitization (including associated infrastructures)
- transnational crime
- human security and health security
- social, political and psychological aspects of security
- recognition and identification (e.g., optical imaging, biometrics, authentication and verification)
- smart surveillance systems
- applications of theoretical frameworks and methodologies (e.g., grounded theory, complexity, network sciences, modelling and simulation)

Together, the high-quality contributions to this series provide a cross-disciplinary overview of forefront research endeavours aiming to make the world a safer place.

The editors encourage prospective authors to correspond with them in advance of submitting a manuscript. Submission of manuscripts should be made to the Editor-in-Chief or one of the Editors.

More information about this series at https://link.springer.com/bookseries/5540

Anna Visvizi · Marek Bodziany

Editors

Artificial Intelligence and Its Contexts

Security, Business and Governance

 Springer

Editors
Anna Visvizi ⓘ
SGH Warsaw School of Economics
Warsaw, Poland

Effat University
Jeddah, Saudi Arabia

Marek Bodziany
The General Kościuszko Military
University of Land Forces
Wrocław, Poland

ISSN 1613-5113   ISSN 2363-9466 (electronic)
Advanced Sciences and Technologies for Security Applications
ISBN 978-3-030-88974-6   ISBN 978-3-030-88972-2 (eBook)
https://doi.org/10.1007/978-3-030-88972-2

This Springer imprint is published by the registered company Springer Nature Switzerland AG
The registered company address is: Gewerbestrasse 11, 6330 Cham, Switzerland

Preface

Artificial intelligence (AI) represents one of the most important developments of our times. The scale of possibilities that AI seems to be promising across fields and domains is vast. And yet the complexity of AI, a technology after all, makes it a subject of both elated expectations and contestation. AI belongs to the group of concepts that while broadly used in the dominant discourse, remain vaguely understood. This applies not only to the mechanisms underlying AI, including artificial neural networks, unsupervised machine learning, deep learning, algorithms, etc., but also to the utilization of AI in our daily lives. The limits of AI's utilization remain unrecognized, and the constraints AI's development faces are rarely talked about. In a similar manner, all too frequently, the discussion on AI fails to highlight the range of technologies, techniques, and applications of which AI is a part. For this reason, in this volume, a case for an AI ecosystem is made. Viewed in this way, AI and its application, as well as the opportunities and risks that it triggers, become a part of a broader context of the (existing and emerging) technologies and new paradigms. This in turn allows a sober discussion on what AI is, what it is not, how it is applied, and what the contingencies and contentious issues relating to AI are. This is what this book is about. Clearly, the content of the book does not exhaust the topic. Nevertheless, it offers a very useful and versatile insight into a range of issues that are central to our understanding of AI and its impact on our societies.

This book would not be possible without the Publisher and the Publishing team, who, once again, proved to be most kind and most professional. In this context, our personal 'thank you' is extended to Dr. Hisako Niko. We would also like to express our gratitude to the reviewers who offered very useful insights into how the book's content could be improved. Finally, we would like to thank the contributing authors for their hard work and patience throughout the lengthy process of editing this volume.

We dedicate this book to the memory of Stanisław Lem (1921–2006), writer and philosopher, whose lucid mind walked the path of human–computer interaction so much earlier than we all did.

Anna Visvizi
Marek Bodziany

Contents

Artificial Intelligence and Its Context: An Introduction

Anna Visvizi⦿ and **Marek Bodziany**⦿

Abstract Considering the scope of novelty that AI brings into politics, economics, and, notably, daily life, it is imperative that it is discussed from as many perspectives as possible. It is imperative that it is discussed in detail, and that the debate on AI is approachable to a variety of stakeholders. This chapter outlines the main themes and issues that have been discussed in chapters that constitute this volume, including those as diverse as the very technical characteristics of AI, AI and politics, ethical questions pertaining to the use of AI in the battlefield, strategic communication (StratCom), military planning, and management and the decision-making process in the domains of business and public administration. The intersection of AI and international trade and national AI strategies from a range of countries around the globe are queried too.

Keywords Artificial intelligence (AI) · Ethics & AI · Artificial neural networks · Deep learning · StratCom

1 Introduction

A certain hype surrounds artificial intelligence (AI): a mixture of elevated hopes and doomsday scenarios as to AI's possible impact on our societies. Central to the discussion are two issues. Viewed in a positive way, AI is seen as the source, almost a panacea, to all possible ills of the current socio-economic model and its ramifications. Viewed in a negative way, AI is considered as the major threat to humankind. In the latter case, it seems that science fiction literature may be the source of inspiration for those who spread these scenarios. However, even if computer science research

A. Visvizi (✉)
Institute of International Studies (ISM), SGH Warsaw School of Economics, Warsaw, Poland
e-mail: avisvi@sgh.waw.pl

Effat College of Business, Effat University, Jeddah, Saudi Arabia

M. Bodziany
Faculty of Security Studies, Military University of Land Forces, Wrocław, Poland
e-mail: m.bodziany@interia.pl

abounds with ever-more sophisticated accounts of what is technically feasible in the domain of AI-enhanced tools and technologies, the bottom line is that the road to these findings' implementation is long and uncertain. At the same time, the general public is scantly aware of this research, and, even less so, of its essence. The emergence and development of AI and related technologies, tools, and techniques, were bound to have triggered multidisciplinary research. Thus, apart from a rich body of research that deals with new ways of utilizing AI in the domains of, say, healthcare, public administration, public transportation, the military and defense, there is a growing body of research that examines AI from the perspective of the humanities, the social sciences, and management research. The key topics that this debate revolves around include, first and foremost, the ethical dimension of AI, but also regulatory barriers, implications for business model innovation, an innovative capacity of the firm, digital transformation of the government and the governance process, and many more. AI and its ecosystem, i.e., a set of paradigms, technologies, approaches, tools, and techniques enabling it and underpinning its development (see Chap. 2 in this volume), are indeed one of the major developments that influence the functioning of our societies. Considering the scope of novelty that AI brings into politics, economics, and daily life, it is imperative that it is discussed from as many perspectives as possible; that it is discussed in detail, and that the debate on AI is approachable for a variety of stakeholders (cf. Polese et al., 2020). The rationale behind this book is a function of this imperative. The objective of this volume is to guide the reader through diverse facets of AI, including its very technical characteristics, the notions of artificial neural networks, algorithms, unsupervised machine learning, deep learning etc. through a range of AI's applications. The following section sheds some light on the book's structure and content.

2   Key Themes and the Structure of the Volume

The chapters included in this volume offer a comprehensive approach to the question of how to conceive of AI and its impact on politics, the economy, and society. The volume is divided into three parts. Part 1 (Chaps. 2–4), titled "AI for good or for what? Contentious issues in and about AI", is devoted exclusively to the perhaps most pressing questions of what AI is, how it may be employed in the policy-making process and how it may be used for security and defense purposes, including combat in diverse theatres of war. Here the question of ethics and the use of AI in combat, including the notion of the "modern" "culture" of killing is upheld. Part 2 (Chaps. 5–7), titled "Security dimensions of AI application", allows the argument to advance and cover issues as diverse as the militarization of AI, the use of AI in strategic communication (StratCom), strategic studies, and in military strategy design. Part 3 (Chaps. 8–10) of the volume, titled "AI, decision making, and the public sector", addresses the question of the value added of AI and machine learning for decision making in the fields of politics and business. In this part of the volume, the searchlight is directed at the notions of data farming, the decision-making process, and AI-supported leadership. In the same section, AI and strategic communication, including disinformation, are discussed, followed by a focus on AI and access to finance. Part 4 of the volume (Chaps. 11–14) consists of case studies covering AI-related developments in the European Union (EU), the United States (US), Saudi Arabia, Portugal, and Poland. The case studies offer an overview of how AI is being employed by the authorities to stimulate growth and development, to promote data and evidence-driven policymaking, and to enable digital governance, international trade, and other related issues (Troisi et al., 2021). The following paragraphs offer a more detailed insight into the content of the chapters included in this volume.

3 Chapter Summary

Chapter 2, titled "Artificial intelligence (AI): explaining, querying, demystifying", and written by Anna Visvizi offers a concise insight into the technical aspects enabling and defining the emergence and development of AI. As the author argues, AI has become a buzzword of today, in several ways reminding us of the concept of "globalization" and the relating debate from two decades earlier. Similar to globalization then, for the greater part of society, AI remains a poorly understood concept, vague, and approached with a fear of the unknown. While AI is hailed as the panacea to all ills of the current socio-economic model and a source of unimaginable opportunities, it is also seen as a source of vital risks and threats relating to safety, security, and the operation of the markets. The objective of this chapter is to explain, query and demystify AI and, by so doing, to highlight the areas and domains that are crucial for AI to develop and serve society at large. To this end, the "dry", i.e., very technical, facets of AI are discussed, and a case for an AI ecosystem is made. Technology-related limitations of AI, as well as possibilities, are outlined briefly. A discussion on AI's impact on the (global) economy and selected policies concludes the chapter.

In Chapter 3, Henrik Skaug Sætra offers a typology of AI applications in politics. As the author argues, politics relates to the most fundamental questions of humans' existence. Modern western societies tend to pride themselves on basing politics on democratic principles, including principles of inclusivity, participation, autonomy, and self-rule. Meanwhile, the continuously improving capabilities of various AI-based systems have led to the use of AI in an increasing number of politically significant settings. AI is now also used in official government decision-making processes, and the political significance of using AI in the political system is the topic of this chapter. The purpose of this chapter is to examine the potential for AI in official political settings, distinguishing between five types of AI used to *support*, *assist*, *alleviate*, *augment,* or *supplant* decision makers. The chapter explores the practical and theoretical potential for AI applications in politics, and the main contribution of the chapter is the development of a typology that can guide the evaluation of the impacts of such applications. AI in politics has the potential to drastically change how politics is performed, and, as politics entails answering some of society's most fundamental questions, the article concludes that the issues here discussed should be subject to public processes, both in politics and society at large.

The focus of Chap. 4, by Marek Bodziany, is directed at the ethical conditions of the use of AI in the modern battlefield. The use of AI in the battlefield causes justified semantic and ethical controversies. Semantic, i.e., relating to meaning, controversies are related to delineating AI used by a human during military operations, e.g., precision weapons capable of independent decision making about the choice of target; and completely autonomous machines, i.e., thinking and deciding autonomously/ independently from the human being. In this broad, and still nascent, field several legal and ethical challenges emerge. This chapter sheds light on the emerging ethical considerations pertinent to the use of robots and other machines relying on unsupervised learning in resolving conflicts and crises of strategic importance for security.

In Chap. 5, Piotr Szczepański, discusses the notions of StratCom, i.e., strategic communication, and communication coherence in the Polish armed forces. StratCom cells are responsible for integrating communication effort and co-ordinating activities in the field of communication capabilities, functions, and tools, including psychological operations (PsyOps), information operations (Info Ops), and military public affairs (MPA). Every day, when observing, for example, image crises, one can see how vital and critical links are caused by incompetent information management. The objective of this chapter is to examine the centrality of StratCom for the success of military operation. Against the backdrop of the Polish Ministry of National Defense, a detailed insight into diverse areas of StratCom is provided. After considering the modern requirements of the information environment, factors influencing the StratCom process are identified. Discussion and recommendations follow.

Chapter 6, by Ferenc Fazekas, dwells on the key issues pertaining to AI and military operations' planning. As the author argues, the key to the success of a military operation rests in its planning. NATO's planning activities are conducted in accordance with the operations' planning process (OPP) and implemented by so-called operations planning groups (OPG). The pace of technological progress has a dramatic impact on the way in which the planning process in the military is conducted. In this context, advances in sensor systems and AI play a major role. The leading militaries already use AI-driven tools, assisting different steps and phases of planning. The same militaries have also already allocated significant budgets to advance AI research and to develop AI assets, such as automated weapon systems and automated planning assistants. However, the real explosion of AI research and AI assets is yet to pragmatize. From a different perspective, in relation to planning, AI is frequently viewed as a potential tool to assist human operators. Notably, advances in AI might lead to the introduction of personal planning assistants, endowed with capacities unmatched by those of human operators. These developments weigh on the OPP. The objective of this chapter is to examine how and to what extent.

In the following Chap. 7, Wojciech Horyń, Marcin Bielewicz, and Adam Joks examine AI-supported decision-making processes in multidomain military operations. The key line of the argument included in the chapter is that modern military operations are conducted in a dynamically changing environment, which necessitates the capacity of instantaneous analysis of huge volumes of data. Due to advances in information and communication technology (ICT), decision makers can be supported in the process of accessing and making use of vast amounts of data and information.

However, due to the instantly available large quantities of data, they are not able to make optimal choices. Technology once again helps to provide several solutions to aid decision makers in making timely and well-thought-out decisions. The operational environment is filled with various devices able to collect, analyze and deliver data in a split second. They are driven by powerful AI engines, use modern simulation software, and use machine-learning algorithms. It seems that the commanders and their staff receive the right and ultimate solutions. However, all that technology is devoid of a social dimension, and their cold analysis is deprived of objectivity and humanity that are essential in making decisions in modern operations. The authors investigate the usefulness and risks of this type of decision support in modern military operations.

In Chap. 8, titled "Artificial intelligence systems in decision-making processes", the author, Marian Kopczewski, argues that the theory and practice of management have long been looking for methods and tools enabling the implementation of management functions in an organization based on a decision-making process that is reliant on modern information technologies. ICT has been successfully supporting management processes in organizations for many years, including in the form of integrated management information systems, decision support systems or systems known as business intelligence. The objective of the chapter is therefore to identify the impact of AI on the decision-making process based on the theory of fuzzy sets and fuzzy logic explaining the ways in which decision systems operate.

Chapter 9, by Gennaro Maione and Giulia Leoni, titled "Artificial intelligence and the public sector: the case of accounting", examines the notion of the impact of AI on public administration. It is argued that the implementation of AI-enhanced systems in the business context offers many benefits. However, in accounting studies, AI is still configured as an almost unexplored frontier. This gap is even more clear for public sector accounting, given the relatively small number of scientific contributions dedicated to the topic. In light of these considerations, the work aims to highlight the limiting factors and the enabling drivers for AI in public sector accounting. To this end, a qualitative approach was applied to study the answers provided by a sample of 45 managers, placed at the head of the accounting office of some Italian municipalities. The questions were prepared in the form of semi-structured interviews, developed by enucleating and adapting the key concepts related to the five attributes that, according to Innovation Diffusion Theory, characterize every innovation process: relative advantage; compatibility; complexity; trialability; and observability. The findings suggest that Italian public sector accounting is experiencing a transition, placing itself halfway between the "early adopter" and the "early majority", that is, in a phase in which current technologies begin to be perceived as outdated and not worthy of further investment, while the process of spreading new AI-based technologies appears interesting but still immature.

Chapter 10, by Giulia Leoni, Francesco Bergamaschi and Gennaro Maione offers an insight into the role and influence of AI at the local government level. Technological advances (from almost unlimited computer power to AI-based systems) combined with managerial information systems may prove significantly valuable under two different areas. On the one hand, business intelligence and business analytics, the latter of which is a layer on which AI commonly builds, may over-come an information overload, absence of cause-effect relationships and a lack of holistic views of the organization. On the other hand, they may improve account-ability while also engaging different stakeholders. However, the diffusion of this approach, especially in the public sector and at a local level, is still both theoretically and practically in its infancy. Thus, this conceptual paper highlights the pivotal role of data-driven decision making and data visualization in local governments, involving the points of view of both accounting scholars and an AI practitioner. The litera-ture review suggests how data analytics may allow the alignment between political programs, strategic goals, and their implementation, fostering the integration between strategic and operational goals. Moreover, data analytics may increase accountability and support the managerial and political decision-making process, increasing their awareness. This chapter calls for new models for public administration managerial decision making, reporting, and organizational culture, which encompass analytics, capable of simplifying data and amplifying their value.

The discussion in Chapter 11, by Jacek Dworzecki and Izabela Nowicka, queries AI in connection with the activities and performance of police formations in Poland. As the authors explain, contemporary challenges in ensuring public safety require the use of ICT-enhanced solutions, including AI and related tools and applications. However, regardless of the effectiveness of police operations carried out with the use of this type of ICT-based tools, questions arise as to the impact of these on civil rights and freedoms. Two facets of this issue exist: i.e., on the one hand, very effective tools of police surveillance exist, e.g., control of electronic correspondence, use of information databases on citizens. These tools enable effective detection of crime and their perpetrators. On the other hand, a negligent use of the same tools may violate the individual's constitutional rights and freedoms. The objective of this chapter is to map and assess the ICT-enhanced solutions employed by the Polish police formations in view of the impact of these tools on civil rights and freedoms.

In Chap. 12, Mikołaj Kugler, examines the US' stance towards AI in relation to defense. As the author argues, the US' adversaries and competitors have recognized this potential and are developing significant initiatives to adopt AI in pursuit of their national security goals, with their investments posing a threat to US' military advantage. US supremacy and leadership do not remain uncontested any longer. With the situation as it is, the US should act to protect its security and advance its competitiveness, leading in the development and adoption of AI defense solutions. Given the above, this chapter will explore and discuss the ways in which the US adopts AI for defense purposes so that it maintains its strategic position and is still capable of advancing security, peace, and stability. It will also provide an overview of AI definitions, an outline of AI's potential impact on national security and the presentation of China's and Russia's approach to encompassing AI-enabled defense capabilities.

In Chap. 13, by Saad Haj Bakry & Bandar A. Al-Saud, Saudi Arabia's National AI Initiative Framework is discussed. AI receives increasing attention due to its potential relating to innovation and socio-economic growth and development. Several countries worldwide initiated national AI initiatives to explore AI for their own benefits, and to gain a competitive edge in AI for future development. The purpose of this chapter is to develop a generic national AI initiative framework that can be useful for countries concerned with planning for their future with AI. Against the backdrop of Bakry's wide-scope and structured STOPE (Strategy, Technology, Organization, People, and Environment) view, the elements of currently available key national AI initiatives are mapped. The resulting framework is then used as a base for proposing an initial AI initiative for Saudi Arabia that considers responding to the requirements of its national development vision known as "KSA Vision 2030". It is hoped that the developed framework and its application will provide a roadmap for developing or improving national AI initiatives in various countries.

The focus of Chap. 14, by Katarzyna Żukrowska, is directed at AI and international trade. As the author observes, AI will have a profound impact on the economy and trade. This chapter examines the scope of changes AI will thus trigger in international trade. This chapter concentrates on the issue as it pertains to the EU member states, the changes on the internal market, the Union's relations with outside partners, and as set within a wider context. The analysis was conducted amid the conditions caused by the slowing economy of 2018 and the onset of the Covid-19 pandemic. Crises often create new conditions and accelerate changes, a theory which is supported by arguments from different fields. This chapter suggests that an acceleration of change took place and was caused by policies applied to overcome the pandemic and by changes in market players' behaviours. This chapter addresses three questions: Will the volume of trade expand or shrink in the very near future? What trends can be expected in different groups of states using the criteria of advancement of development? And, can AI be considered as stimulation accelerating changes in trade?

4 Who Should Read This Book and Why?

There is a wealth of books and journal articles that address the complex notion of AI in the modern world (Russel & Norvig, 2020; Siskos & Choo, 2020; Lytras & Visvizi, 2021; Visvizi & Lytras, 2019). Frequently, but not always, the focus of the discussion that unfolds in these books is either highly technical, i.e., focused on a very narrow field of application, or dominated by philosophical considerations pertaining to AI. These include, for instance, the question of human agency in the machine era. Recognizing the value of these publications, this volume avoids the pitfall of being over-technical, or claiming to explain technology to amateurs, or raising a false pretense of explaining … magic. The chapters included in this volume elaborate on technology as much as it is necessary for the reader to understand the concepts and the mechanisms behind the issues and processes that this book addresses, i.e., AI and its ecosystem (cf. Visvizi, 2021). For this reason, this volume is not overwhelming to non-technical readers. Written by practitioners and researchers, it offers a hands-on critical insight into a range of issues and developments in the socio-political and economic realms that are influenced by AI. The discussion in this volume engages with two critical dimensions in which the onset of AI is paramount: security and defense; and the decision-making process in politics, business, and the economy. These two axles of the discussion are complemented by a selection of regional case studies. This combination of conceptual and empirical insights into AI will allow the reader to understand how the rapid development of AI is playing out in the European Union, the US, the GCC, and Saudi Arabia, and—more broadly—in international trade.

In brief, this volume will be of great value to all those (academics, practitioners, and policy makers) who have recognized that AI is altering the ways in which the world works and still seek to understand how it happens. To this end, the book employs practitioners' insights and filters them through the existing debate and the nascent literature in the field, to offer a primary reference point for researchers, academics, and students, including those at the executive level of education.

References

Kashef, M., Visvizi, A., & Troisi, O. (2021). Smart city as a smart service system: Human-computer interaction and smart city surveillance systems. *Computers in Human Behavior, 2021,* 106923. https://doi.org/10.1016/j.chb.2021.106923

Lytras, M. D., & Visvizi, A. (Eds.). (2021). *Artificial intelligence and cognitive computing: Methods, technologies, systems, applications and policy making.* MDPI Publishing. ISBN 978-3-0365-1161-0 (Hbk), ISBN 978-3-0365-1160-3 (PDF), https://doi.org/10.3390/books978-3-0365-1160-3

Polese, F., Troisi, O., Grimaldi, M., & Sirianni, C. (2020). *Conceptualizing social change, between value co-creation and ecosystems innovation* (p. 203). Festschrift für Universitätsprofessor Dr. Dr. hc Helge Löbler.

Sarirete, A., Balfagih, Z., Brahimi, T., Lytras, M. D., & Visvizi, A. (2021). Artificial intelligence and machine learning research: Towards digital transformation at a global scale. *Journal of Ambient Intelligence and Humanized Computing.* https://doi.org/10.1007/s12652-021-03168-y

Sikos, L. F., & Choo, K. K. R. (Eds.). (2020). *Data science in cybersecurity and cyberthreat intelligence.* Springer. https://www.springer.com/gp/book/9783030387877

Russell, S., & Norvig, P. (2020). *Artificial intelligence: A modern approach* (3rd Ed.). Pearson. https://www.amazon.com/Artificial-Intelligence-Modern-Approach-3rd/dp/0136042597

Troisi, O., Visvizi, A., & Grimaldi, M. (2021). The different shades of innovation emergence in smart service systems: The case of Italian cluster for aerospace technology. *Journal of Business & Industrial Marketing,* ahead-of-print. https://doi.org/10.1108/JBIM-02-2020-0091

Visvizi, A. (2021) Artificial intelligence (AI): Explaining, querying, demystifying. In A. Visvizi, & M. Bodziany (Eds.), *Artificial intelligence and its context—Security, business and governance.* Springer.

Visvizi, A., & Lytras, M.D. (Eds.). (2019). *Politics and technology in the post-truth Era.* Emerald Publishing. ISBN 9781787569843, https://books.emeraldinsight.com/page/detail/Politics-and-Technology-in-the-PostTruth-Era/?K=978178756984

Anna Visvizi Ph.D., Associate Professor, SGH Warsaw School of Economics, and Visiting Scholar at Effat University. Jeddah. An economist and political scientist, editor, researcher and political consultant, Prof. Visvizi has extensive experience in academia, think-tank and government sectors in Europe and the US. Her expertise covers issues pertinent to the intersection of politics, economics and ICT. This translates in her research and advisory roles in the area of AI and geopolitics, smart cities and smart villages, knowledge & innovation management, and technology diffusion, especially with regards to the EU and BRI. ORCID: 0000-0003-3240-3771 Email: avisvizi@gmail.com

Marek Bodziany Ph.D. in Sociology, Associate Professor, Colonel of the Polish Armed Forces, Deputy Dean of the Faculty of Security Sciences at the Military University of Land Forces, Chief Editor of the AWL Publishing House. Research areas: multiculturalism, education, culture sociology, ethnical and cultural conflicts, wars and social crises, methodology, migration and social change. ORCID: 0000-0001-8030-3383 Email: m.bodziany@interia.pl

AI for Good or for What? Contentious Issues In and About AI

Artificial Intelligence (AI): Explaining, Querying, Demystifying

Anna Visvizi ⓘ

Abstract Artificial intelligence (AI) is a buzzword today, reminding us of the concept of "globalization" and the relating debate two decades ago. As with globalization then, for the greater part of society, AI remains a concept poorly understood, vague, and approached with fear of the unknown. While AI is hailed as the panacea to all the ills of the prevailing socio-economic model and a source of unimaginable opportunities, it is also seen as a source of substantial risks and threats to safety, security, and the operation of the markets. The objective of this chapter is to explain, query and demystify AI and by so doing to highlight the areas and domains that are crucial for AI to develop and serve society at large. To this end, the "dry", i.e., quite technical, facets of AI are discussed, and a case for an AI ecosystem is made. Technology-related limitations of AI, as well as possibilities, are outlined briefly. An overview of AI's implications for the (global) economy and selected policies follows. The ethical concerns are discussed in the concluding section.

Keywords Artificial intelligence (AI) · AI ecosystem · International economy

1 Introduction

Artificial intelligence (AI) has become a buzzword of today, reminiscent of the concept of "globalization" and the relating debate on globalization two decades or so earlier (cf. James & Steger, 2014; Rosamond, 2003; Turenne Slojander, 1996). As in the case of globalization and the discourse surrounding it (however defined!), AI remains a concept both powerful and yet insufficiently explained, explored or understood by large sectors of society. AI is hailed by many as the panacea to all the ills of the current socio-economic model as well as a source of unimaginable opportunities (cf. Zhuang et al., 2017; Eager et al., 2020; Floridi, 2020). However, AI, especially in popular discourse, is also seen as a source of risks and threats. Clearly, it is too early to

A. Visvizi (✉)
Institute of International Studies (ISM), SGH Warsaw School of Economics, Warsaw, Poland
e-mail: avisvi@sgh.waw.pl

Effat College of Business, Effat University, Jeddah, Saudi Arabia

A. Visvizi and M. Bodziany (eds.), *Artificial Intelligence and Its Contexts*, Advanced Sciences and Technologies for Security Applications,
https://doi.org/10.1007/978-3-030-88972-2_2

imagine and understand all the possible implications, positive and negative, that the use of AI-enhanced solutions may trigger for our societies, including their politics, economics, business sectors, competition, and innovation, as well as education and healthcare provision etc. Therefore, a thorough and unbiased conversation, including academics, researchers, and practitioners, on AI and AI-related topics is necessary. This chapter and thus the entire volume respond to this plea by offering an insight into diverse aspects of AI and its use across wide-ranging issues and domains, including security and military affairs, the decision-making process, and public administration. Interestingly, amidst the frequently alarming media accounts of AI and the alleged existential risks and threats to humanity that it bears, AI-based solutions, including services and applications, have been present in our lives for some time. Consider the autocorrect feature in smart phones, and how it picks the key language you use daily, and how it learns throughout the process to recognize which language you are using at a given moment when sending your messages or typing your emails. However, AI is much more than language and speech recognition applications in the smart phone. AI and AI-based techniques and applications, and, for example, AI-based solutions, have relevance across very many issues and domains. At present, when advances in AI and related technologies have been transferred from the field of research to application, the challenge is to create appropriate, resilient and flexible regulatory frameworks to ensure that the risks and threats that AI holds for our societies can be preempted and addressed, whilst ensuring that the development of AI and, thus, innovation centered on AI and its ecosystem are not suppressed. To address these complex issues, a thorough understanding of the nature of AI, its limitations and potential is necessary. This chapter aims to provide this.

The discussion in this chapter is structured as follows. In the next two sections, the "dry" i.e., quite technical, facets of AI are discussed. Here a case for an AI ecosystem is made and the notions of artificial neural networks and unsupervised machine learning are discussed. In what follows, technology-related limitations of AI as well as possibilities are outlined briefly. An overview of AI's implications for the (global) economy and selected policies follows. The ethical concerns are discussed in the concluding section.

2   AI and Its Ecosystem

The popular understanding of AI and, for that matter, the key lines of the popular narrative on AI portray AI as a sort of crystal ball capable of foreseeing the future, a super-computer capable of resolving the most complex problems, a super-weapon of total destruction, and perhaps also as a super-machine waiting to replace human beings. This conception of AI, characterized by fear of the unknown, has been famously captured in Lem's (1964, 1990) visionary robots' fables, the thrust of which concerned the big question of the human–computer (aka robot) interaction. The onset of AI and, perhaps most importantly, the machine-learning-mediated possibility for people to enter a form of interaction with a computer/machine revives the basic question of how to conceive of the absolutely new form of communication, i.e., with a non-human. To be clear, the objective of this chapter and thus, also of the entire volume, is to showcase that, regardless of the potential inherent in AI and related technologies, AI should not be reified or, in other words, considered as ontologically distinct from the human being, even if valid fears exist that AI might replace the human (Archer, 2021; Rakowski et al., 2021; Sætra, 2020; Fiske et al., 2019). To understand why, it is necessary to shed light on the very essence of AI.

Contrary to popular (mis)understanding, AI is best conceived as an ecosystem of sophisticated technologies, techniques, and applications. Their emergence is a function, first and foremost, of the development of ever more efficient processors capable of handling ever more complex calculations in an ever-shorter time span, and thus delivering greater power necessary for our computers to operate and perform the tasks we require (cf. Khan et al., 2021). Central to the emergence of AI is the development of neural networks, i.e., several layers of interconnected processors that, similarly to neurons in a human brain, have the capacity to share impulses and, thus, data.

The emergence of AI is directly related to the onset of the big data paradigm in computer science, and thus the ability to process huge data sets in a short period of time, a task impossible before (Farami et al., 2021; Visvizi et al., 2021). The emergence of the big data paradigm changed the way data is conceived and valued. That is, now, huge datasets become a source of invaluable information across domains (Lytras et al., 2021). Yet the capacity to process the phenomenally huge and complex sets of data is only possible due to the emergence and application of such software technologies as Apache Hadoop and other. These technologies enable, on the one hand, a different and a more efficient way of processing data, and, on the other hand, distributed processing of large data sets across clusters of computers by means of simple programming models.

In other words, as Fig. 1 outlines, any discussion on AI needs to be based on the recognition that it is an AI ecosystem rather than a single super-technology. Specifically, the emergence, development and utilization of AI are driven by three

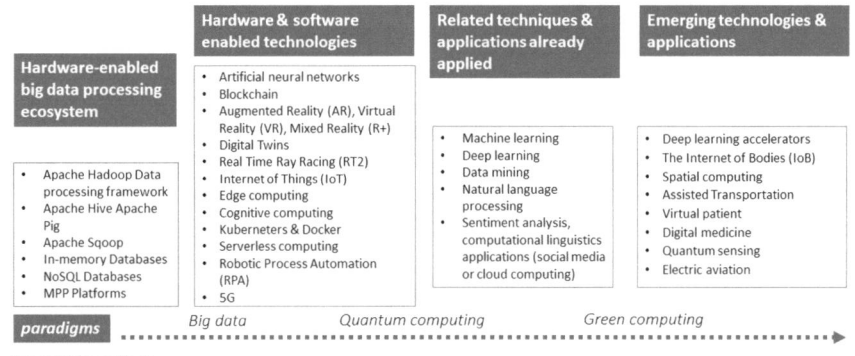

Source: The author.

Fig. 1 AI and its ecosystem. *Source* The Author

major paradigms that shape the debate in the discipline of computer science, i.e., the big data paradigm, the quantum computing research, and green computing (Hu et al., 2021; Piattini et al., 2020; Lytras et al., 2021). These three broad debates, representing interconnected but paradigmatic shifts, shape the developments in the discipline and in the field. As such, they are consequential for the remaining items in the AI ecosystem, such as the (i) hardware-enabled big data processing frameworks and open-source software utilities, (ii) hardware- and-software enabled technologies; (iii) related techniques and applications already in application; (iv) emerging technologies and applications. Notably, the landscape of techniques, technologies, and applications, all fostered by advances in the three basic paradigms that shape advances in information and communication technology (ICT) today, changes rapidly. Thus, stories that hold truth today, are subject to contestation and revalidation tomorrow. Knowledge is emancipatory after all (Archer, 1998; Bhaskar, 1975).

3   AI and the Neural Networks: From Supervised to Unsupervised Learning

To put it simply, AI is the—enhanced by super-processors—process of grouping, structuring and analyzing huge sets of data, especially of unstructured data, i.e., the kind of data that has not been previously "tagged". Notably, AI is not a new term. Already in the 1950s, attempts were made to utilize the learning potential of computers of that period. At that time, however, the focus of the activity was directed at so-called supervised machine learning. The latter employs algorithms that require structured or tagged data. The machine learning process consists in this case of the classification of data based on a key, i.e., a key included in the algorithm, and drawing conclusions based on the classic probability tree. Even if supervised machine learning is still in use in some domains today (cf. Jiang et al., 2020), it displays fundamental weaknesses. Its efficiency is directly related to the possibility and ability of the human, in this case the system manager, to label or tag the data. Considering that the data sets are huge and that the supply of data is massive and continuous, tagging (or labelling) data is time-consuming, requires a lot of skill and thus is very expensive.

In contrast to supervised machine learning, unsupervised learning—and it is the essence of the concept of AI—allows unstructured data to be fed into the system and subsequently examined (cf. Alloghani et al, 2020; Rajoub, 2020). To explain the meaning of unsupervised learning, it is necessary to highlight three basic terms, i.e., artificial neural networks, algorithms, and deep learning. The artificial neural networks consist of an immense number of connected processors operating at the same time, which as per the design—are supposed to mirror the functioning of the human brain. Each of the processors that builds a part of each of the neural networks has access to the local memory and is fed by, on the one hand, huge data sets and, on the other hand, information about the relationships among the data. The neural networks work as a system, i.e., they may overlap—thus the talk of layers of the artificial neural networks—and co-operate.

The content and quality of the algorithms define the ways in which the neural networks and the layers operate and co-operate. A given algorithm defining the functioning of a neural network is responsible for the process of (i) recognizing the characteristics of data available in the system, a process called feature introspection; (ii) identifying similarities among chunks of data, a process called similarities detection; and (iii) grouping these unstructured data into clusters. In other words, algorithms establish the relationships between and among data, e.g., vanilla ice-cream (taste), and not, say, red ice-crem (color) (Strickland, 2019). Taking into account that neural networks can overlap, i.e. it is possible to use several layers of neural networks, and an algorithm enables the exchange of data/information among them, it is possible to talk about the learning process of a neural network. However, for a neural network to learn, the algorithm and the software that operationalize it must demonstrate to the neural network how it has to respond to external cues, e.g., to data introduced by the user.

AI is based on the utilization of the outcomes of this kind of multilayered co-operation of complex neural networks. Precisely this feature of unsupervised learning, consisting of the multilayered co-operation among neural networks, defines the thrust of what is otherwise referred to as "deep learning". Viewed in this way, the process of deep learning is feasible due to specific algorithms and advanced processors. Under these conditions, in a complex multilayered neural network, based on co-operation among specific layers of the neural network, the following processes take place: introspection, similarities' detection, clusters' grouping, and, finally, delivering an answer to a question included in the algorithm.

Considering that the number of data available in the system increases, the artificial neural networks have to expand, the power/capacity of processors has to increase, the number of relationships between and among chunks of data becomes denser, and thus the learning capacity of a given neural network increases, always in line with the logic defined in the algorithm that defines its operation. It remains to be seen how and in which direction artificial neural networks will develop. Now, irrespective of the fact that AI tends to be viewed as an embodiment of unsupervised learning, and therefore also as a form of deep learning, the classic rules deriving from the science of probability are also at play in the process of the functioning of neural networks. Having said that, it is necessary to dwell on the limitations, possibilities and risks inherent in neural networks.

4 Limitations, Possibilities and Risks Inherent in Neural Networks

AI is directly correlated with the processing power of respective processors as well as with the quality of the algorithms that define a neural network's operation and co-operation. From a different angle, regardless of arguments suggesting that neural networks can do everything, neural networks are not intelligent enough. In other words, a neural network cannot make decisions about and create the logic of thinking that subsequently will allow it to connect data in line with certain rules. The latter are quintessential for the processes of introspection, similarities detection, and clustering, i.e., for the process of building a logical whole out of a myriad of pieces. Notably, attempts to teach neural networks, e.g., by means of generating cookbooks led to quite anecdotal combinations such as "Take 250 gr of bones or fresh bread … ", etc. In other words, the problem, or a challenge, rests in the fact that a neural network repeatedly committed mistakes suggesting it did not have any memory at all (Strickland, 2019). To put it differently, artificial neural networks can only work forward, i.e. in a range from T_0 to T_1. They cannot predict the future. Artificial neural networks can support real-time tasks and commands.

Another, by now classic, example that was meant to hail the success of AI relates to AlphaGo, i.e. the software and a game from the company DeepMind (cf. Granter et al., 2017; Li & Du, 2018; Holcomb et al., 2018). Go is a traditional, highly complex, Chinese board game, and its history goes back 3000 years. Attempts to utilize supervised machine learning to develop a software version of the game allowed for an amateur-level version of the game (cf. Lu et al., 2020). This is because the number of possible moves, the number of their combinations, and the evaluation of their relative values for the follow-up of the game were far more than the possibilities inherent in the traditional probability "trees". The advent of neural networks and so, also, the onset of unsupervised learning made it possible to bypass this problem. That is, the software AlphaGo, developed by Deepmind (and purchased by Google for USD 500 m in 2014, (cf. Shu, 2014), employs an advanced probability equation to filter the layers of the neural network that is structured according to such values/functions as "politics", "values" etc., i.e., items specific for the game (cf. Binder, 2021).

An interesting discovery relating to the AlphaGo is that the software was unable to define what "human being" is. Even more so, the software was unable to discern whether it was competing with a machine or with a human being. To put it differently, at present, the algorithms underpinning the functioning of neural networks demonstrate great precision whenever the problem/task at hand is narrow and very well defined. Thus, the more specific a given problem, the smarter AI appears to be. The opposite is also true. For instance, should a given algorithm be designed to generate pictures and, more specifically, pictures of birds, it will not be able to generate a picture of any other animal.

Discussion of possibilities inherent in AI needs to take into account the parallel advances in sophisticated technologies, applications, and techniques related to the process of the handling of big data. This includes the highly technical issues already at the level of the processor and its speed, as well as exchange of information between and among processors, etc. These factors, as mentioned earlier, directly influence the possibilities relating to deep learning, and especially the pace of deep learning and the condensation of information that it may lead to. Possibly, if only the processors' capacity would increase as fast as the increasing quantity of data they are fed with, one could argue that the potential inherent in AI was unlimited. At this point, it is worth mentioning the case of deep learning accelerators (see also Fig. 1), i.e., a technology that seeks to mitigate some of the technical limitations specific to the limited overall speed of the existing processors. In this context, it is also important to highlight the case of quantum computing, i.e. an emerging paradigm that revolutionizes the way in which the very process of computing is conducted and that, at the same time and for precisely this reason, a myriad of possibilities and applications across fields and domains open up (Hassija et al., 2020; Lee, 2020; Outeiral et al., 2021).

With regards to AI's limitations, several barriers exist that limit the feasibility of its utilization. These barriers can be identified at diverse levels and in several overlapping domains and issues specific to contemporary politics, society, and the economy. These barriers may include: digital illiteracy; lack of awareness and expertise regarding the use of AI-enhanced tools; lack of resources to purchase AI-based tools and solutions; and a lack of, or insufficient, infrastructure that would enable rightful access to data and their efficient use. The poor quality of data, i.e. data in an improper format, constitutes another obstacle to efficient AI utilization (cf. OECD, 2015; Hatani, 2020). As shall be discussed in the following section, obstacles and limitations to the use of AI-based solutions rest also in the regulatory field, including questions of privacy, personal data protection, safety and security, grey areas of regulations, as well as regulatory barriers, insufficient complementarity of regulation in the field of AI in the international context and many others. Considering the breadth of the discussion focused on contentious issues pertaining to the emergence, development and application of AI in modern world, the following section will merely outline the key areas of concern. The content of the volume (cf. Visvizi & Bodziany, 2021) will serve as a necessary addition to the brief introduction included in this chapter.

5   AI and the Society: Focus on the (Global) Economy and Selected Policies

From the perspective of what is technically feasible, the onset of AI and AI-based solutions, applications and techniques is a source of a great number of opportunities. These opportunities may be seen through several interpretive lenses. For example, viewed from the perspective of the global economy, it may be assumed that further automation and digitalization of production, this time employing AI-based solutions, would allow an aggregate global productivity increase. This would be due to the work substitution effect: the rise in innovativeness and innovation and the adoption of new technologies; increased competition; diversification of products; optimization of supply-chains as well as of global value chains, etc. From a different point of view, AI is a source of great hope in the field of healthcare in that the inroads of AI-based techniques and applications, always in connection with other technologies, revolutionize the way in which data is handled for the purposes of early detection and diagnosis of diseases, treatment, medicine development, the analysis of archived patients' records, surgery and so much more (Lytras et al., 2021). AI, as the content of this volume amply demonstrates, is employed in public administration, the services sector, marketing (Huang & Rust, 2021; van Veenstra et al., 2020; Loukis et al., 2020) and, of course, in the field of the military. Viewed in this way, the hopes, prospects, and opportunities that AI-based solutions bring to our society are immense. However, as is the case with any new technology or an ecosystem of technologies, tools, techniques and applications, downsides always exist, and a period of adjustment is necessary. Consider the case of the automobile in the late nineteenth century (Berger, 2001), and the regulatory frameworks that we continue to put in place to strike a sound balance between the benefits of using a car and the externalities that the aggregate use of vehicles creates for the environment.

The case of AI proves that, even if its value added and its potential are immense, our societies, but perhaps more precisely the regulatory frameworks and the governance structures in place, are not fit enough to accommodate not only the plethora of regulatory challenges ahead but also the variety of resulting, and yet still unaccounted for, policy-making considerations that will have to be addressed (cf. Ulnicane et al., 2021). Substantial effort has been undertaken at the global (G20), international (United Nations, the Organization for Economic Co-operation and Development), regional (the European Union (EU)), and national levels to address the challenge (e.g., Chatterjee, 2020). In some ways, as more critical voices outline, the "race to AI" has turned into a global "race to AI regulation" (Smuha, 2021), in which a country's capacity to influence international standards and regulations in the field of AI turns into competitive advantage (Dexe & Franke, 2020). Consider the case of the EU, a global player after all. 290 AI policy initiatives devised by the EU member states exist. The EU, especially the European Commission, makes an explicit attempt to promote a broad approach to AI (European Commission, 2020; European Commission, 2021). AI is thus seen as a tool, on the one hand, to help the EU compete globally and, on the other hand, to—even if this notion is quite implicit—allow the EU to reclaim its position as the "normative power of Europe" (Manners, 2002). In relation to the first objective, the European Commission places emphasis on innovative capacity at large, and in this context on the use of data and developing "critical computer capacity". In relation to the second objective, the European Commission's narrative is filled with references to "AI for the people", "AI as a source of good", and to "trustworthy AI" (European Commission, 2018).

It is difficult not to be critical of the approach to AI that the European Commission promotes. It seems like too little, too late, and not comprehensive enough. In other words, the Commission's approach to AI and to the prospect of reaping the benefits of AI, resembles hand-steering rather than anything else. The point is that too much emphasis is placed on direct-support measures for, say, research and investment in AI, at the expense of letting the market mechanisms play their due part as well. This is certainly a nod to all those who deal with economic policy in specific EU member states and with economic policy coordination at the euro-area level. The notions of the EU competition policy and industrial policy complete the picture. Full stop. The point is that hand-steering alone will not allow small- and medium-sized enterprises (SMEs) in the EU, or elsewhere in the world, to pick up and adopt AI-based tools and solutions in their daily activities (cf. Troisi et al., 2021). For this, tax incentives and flexible labor markets are needed to enable SMEs to acquire talent, which is rare and highly expensive, and invest in AI-based solutions.

The downside of this story, as many observers note (cf. Naudé, 2021), is that, since AI-based tools are key to digital platforms and digital platform economy, and data is the key source of advantage, as the case of Google, Apple, Facebook, Amazon, and Alibaba (GAFAA) illustrates, we witness a major distortion of competition these days. The thrust of the problem is that the barriers to entry are so high that companies like GAFAA lack natural competitors. From a different perspective, though, the case of Microsoft and the way it was dealt with in the context of the EU competition policy more than a decade ago demonstrates that a thin line divides attempts to defend fair competition from those to create disincentives for innovation (Larouche, 2009; Manne & Wright, 2010). In brief, the case of the EU is illustrative of the scale of the challenge that the inroads of AI create for the economy and for the regulators.

Looking at this issue from a different perspective, it is necessary to mention the "first-mover advantage", i.e., a situation when companies that were first in the field are capable of maintaining their dominant position, at the expense of other market players (cf. Park, 2021). This is already happening, cf. GAFAA. Consider, however, the global implications of the same process. That is, how will the inroads of AI shape and reshape the Global North-Global South relationship? In the context of the debate on the Global South, the "first-mover advantage" thesis suggests that companies from the Global North, i.e., current leaders, might be able to capture the opportunities first, effectively denying them to actors in the Global South. This might lead to exclusion and reproduction of the old global dependency patterns. Of course, the case of China adds to the complexity of the issue. In brief, with reference to AI and its impact on the global economy, time and coordination are crucial. It remains to be seen whether a consensus on these issues will be reached at G20 any time soon.

6 By Means of Conclusion

While the economy has been centrally featured in the discussion in the previous section, this chapter would be incomplete if some of the ethical considerations pertaining to the development and adoption of AI-based solutions were not mentioned. Several exist. The thrust of the concern is twofold, i.e., to what extent will the computer, the machine, be able to operate independently of the human being and regardless of the human being's express consent, and how can it be ensured that it will be a force for good? There is a growing body of literature that deals with the question of ethics and AI (Jobin et al., 2019; Coeckelbergh, 2020; Vesnic-Alujevic et al., 2020; Roberts et al., 2021; Ryan & Stahl, 2021; Rakowski et al., 2021; Neubert & Montañez, 2020). International organizations, such as the UN, the EU, the World Health Organization, as well as national governments (Dexe & Franke, 2020) are central in shaping and developing the debate. A very useful typology of ethical considerations that the debate on AI and ethics revolves around has been introduced by Jobin et al. (2019) and further elaborated by Ryan and Stahl (2021,

p. 65). In brief, the following ethical considerations are key for the debate in AI in our societies: transparency, justice and fairness, non-maleficence, responsibility, privacy, beneficence, freedom and autonomy, trust and trustworthiness, sustainability, dignity, and solidarity. What are the practical implications thereof? As the chapters included in this volume demonstrate, the use of AI-enhanced tools in the battlefield is one of the most contentious issues in this respect. This certainly applies to unmanned aerial vehicles (UAVs), or simply drones, and warfare. Consider non-maleficence. The case of facial features' recognition in China raised very serious ethical concerns. Consider justice and fairness, privacy, freedom and autonomy and so on. The question of who, under which conditions and for what purpose, has the right to collect, or who gains access to, our personal data is another case that highlights sensitive issues surrounding AI-based tools and techniques. The General Data Protection Regulation (GDPR) implemented in the EU since 2018 is one of the strictest regulatory responses to this issue in the world.

The conversation on AI and ethics leads to another question, i.e., prospective directions of growth and development of AI and AI-based tools. In this context, it is necessary to return to the initial point that AI is bound to fundamentally change the relationship between the human being and the machine. This time, it is not a literary question. This is because current and future users of technology will have to learn a form of interaction with a non-human. Consider the case of education and technology-enhanced learning based on AI and cognitive computing. By means of enhancing the user's experience, and thus amplifying the teaching and learning process (Lytras et al., 2018; Visvizi et al., 2020), already today, young students should acquire the necessary skills and ethical stance that would allow them to shape the future human–computer relationship in a responsible manner.

Another issue that should be mentioned here relates to the question of the impact of AI on inequality, exclusion, and, from a different perspective, non-growth. It is possible to imagine that the emerging economic system, to a large extent based on digital platforms (Alahmadi et al., 2020; Malik et al., 2021), will also lead to the consolidation of a new system of social relations, of wealth, and wellbeing, both locally and globally. Depending on the way of reading it, this scenario may be devastatingly daunting. To prevent it from happing, transparency understood as explainability, explicability, understandability, interpretability, communication, disclosure, and showing, is the key issue in the debate on AI. This chapter has sought to do just that.

References

Alahmadi, D., Babour, A., Saeedi, K., & Visvizi, A. (2020). Ensuring inclusion and diversity in research and research output: A case for a language-sensitive NLP crowdsourcing platform. *Applied Sciences, 10*, 6216. https://doi.org/10.3390/app10186216

Alloghani, M., Al-Jumeily, D., Mustafina, J., Hussain, A., & Aljaaf, A. J. (2020). A systematic review on supervised and unsupervised machine learning algorithms for data science. In M. Berry, A.

Mohamed, & B. Yap (Eds.), *Supervised and unsupervised learning for data science. Unsupervised and semi-supervised learning.* Springer. https://doi.org/10.1007/978-3-030-22475-2_1.

Archer, M. S. (2021). Friendship between human beings and AI Robots? In J. von Braun, M. S. Archer, G. M. Reichberg, & M. S. Sorondo (Eds.), *Robotics, AI, and humanity.* Springer. https://doi.org/10.1007/978-3-030-54173-6_15

Archer, M. S. (1998). *Critical realism: Essential readings.* Routledge.

Berger, M. I. (2001). *The automobile in American history and culture: A reference guide.* Greenwood Publishing Group. ISBN 978-0313245589.

Bhaskar, R. (1975). *A realist theory of science.* Books.

Binder, W. (2021). AlphaGo's deep play: Technological breakthrough as social drama. In J. Roberge, & M. Castelle (Eds.), *The cultural life of machine learning.* Palgrave Macmillan. https://doi.org/10.1007/978-3-030-56286-1_6.

Park, C. (2021). Different determinants affecting first mover advantage and late mover advantage in a smartphone market: A comparative analysis of Apple iPhone and Samsung Galaxy. *Technology Analysis and Strategic Management.* https://doi.org/10.1080/09537325.2021.1895104

Chatterjee, S. (2020). AI strategy of India: Policy framework, adoption challenges and actions for government. *Transforming Government: People, Process and Policy, 14*(5), 757–775. https://doi.org/10.1108/TG-05-2019-0031

Coeckelbergh, M. (2020). *AI ethics.* MIT Press.

Dexe, J., & Franke, U. (2020). Nordic lights? National AI policies for doing well by doing good. *Journal of Cyber Policy, 5*(3), 332–349. https://doi.org/10.1080/23738871.2020.1856160

Eager, J., Whittle, M., Smit, J., Cacciaguerra, G., & Lale-Demoz, E. (2020). Opportunities of artificial intelligence, report, policy department for economic, scientific and quality of life policies, directorate-general for internal policies, PE 652 713—June 2020. European Parliament.

Eskak, E., & Salma, I. R. (2021). Utilization of artificial intelligence for the industry of craft (November 5, 2020). In *Proceedings of the 4th International Symposium of Arts, Crafts & Design in South East Asia (ARCADESA).* Available at SSRN https://ssrn.com/abstract=3807689 or https://doi.org/10.2139/ssrn.3807689.

European Commission. (2018). Draft ethics guidelines for trustworthy AI, working document for stakeholders' consultation. In *The European Commission's High-level Expert Group on Artificial Intelligence.* European Commission. https://digital-strategy.ec.europa.eu/en/library/draft-ethics-guidelines-trustworthy-ai

European Commission. (2020). White paper on artificial intelligence—A European approach to excellence and trust. Brussels, 19.2.2020, COM(2020) 65 final.

European Commission. (2021). Coordinated plan on artificial intelligence 2021 review, annexes to the communication from the commission to the European Parliament, the European Council, the Council, the European Economic and Social Committee and the Committee of the Regions, Fostering a European approach to Artificial Intelligence, Brussels, 21.4.2021, COM(2021) 205 final.

Fararni, K. A., Nafis, F., Aghoutane, B., Yahyaouy, A., Riffi, J., & Sabri, A. (2021). Hybrid recommender system for tourism based on big data and AI: A conceptual framework. *Big Data Mining and Analytics, 4*(1), 47–55. https://doi.org/10.26599/BDMA.2020.9020015

Fiske, A., Henningsen, P., & Buyx, A. (2019). Your robot therapist will see you now: Ethical implications of embodied artificial intelligence in psychiatry, psychology, and psychotherapy. *Journal of Medical Internet Research, 21*(5), e13216. https://doi.org/10.2196/13216.PMID:310 94356;PMCID:PMC6532335

Floridi, L. (2020). AI and its new winter: From myths to realities. *Philosophy and Technology, 33*, 1–3. https://doi.org/10.1007/s13347-020-00396-6

Floridi, L., Cowls, J., Beltrametti, M., et al. (2018). AI4People—An ethical framework for a good AI society: Opportunities, risks, principles, and recommendations. *Minds and Machines, 28*, 689–707. https://doi.org/10.1007/s11023-018-9482-5

Granter, S. R., Beck, A. H., Papke, & D. J. (2017). AlphaGo, deep learning, and the future of the human microscopist. *Archives of Pathology & Laboratory Medicine, 141*(5), 619–621. https://doi.org/10.5858/arpa.2016-0471-ED

Hassija, V., Chamola, V., Saxena, V., Chanana, V., Parashari, P., Mumtaz, S., & Guizani, M. (2020). Present landscape of quantum computing. *IET Quantum Communication, 1*, 42–48. https://doi.org/10.1049/iet-qtc.2020.0027

Hatani, F. (2020). Artificial Intelligence in Japan: Policy, prospects, and obstacles in the automotive industry. In A. Khare, H. Ishikura, & W. Baber (Eds.), *Transforming Japanese business. Future of business and finance*. Springer. https://doi.org/10.1007/978-981-15-0327-6_15.

Holcomb, S. D., Porter, W. K., Ault, S. V., Mao, G., & Wang, J. (2018). Overview on deep mind and its AlphaGo Zero AI. In *Proceedings of the 2018 International Conference on Big Data and Education (ICBDE'18)* (pp. 67–71). Association for Computing Machinery. https://doi.org/10.1145/3206157.3206174

Hu, N., Tian, Z., Du, X., Guizani, N., & Zhu, Z. (2021). Deep-Green: A dispersed energy-efficiency computing paradigm for green industrial IoT. *IEEE Transactions on Green Communications and Networking, 5*(2), 750–764. https://doi.org/10.1109/TGCN.2021.3064683

Huang, M.-H., & Rust, R. T. (2021). Engaged to a robot? The role of AI in service. *Journal of Service Research, 24*(1), 30–41. https://doi.org/10.1177/1094670520902266

James, P., & Steger, M. B. (2014). A genealogy of 'Globalization': The career of a concept. *Globalizations, 11*(4), 417–434. https://doi.org/10.1080/14747731.2014.951186

Jiang, T., Gradus, J. L., & Rosellini, A. J. (2020). Supervised machine learning: A brief primer. *Behavior Therapy, 51*(5), 675–687. https://doi.org/10.1016/j.beth.2020.05.002

Jobin, A., Ienca, M., & Vayena, E. (2019). The global landscape of AI ethics guidelines. *Nature Machine Intelligence, 1*(9), 389–399. Available at https://doi.org/10.1038/s42256-019-0088-2

Kashef, M., Visvizi, A., & Troisi, O. (2021). Smart city as a smart service system: Human-computer interaction and smart city surveillance systems. *Computers in Human Behavior, 2021*, 106923. https://doi.org/10.1016/j.chb.2021.106923

Khan, F. H., Pasha, M. A., & Masud, S. (2021). Advancements in microprocessor architecture for ubiquitous AI—An overview on history, evolution, and upcoming challenges in AI implementation. *Micromachines, 12*, 665. https://doi.org/10.3390/mi12060665

Larouche, P. (2009). The European microsoft case at the crossroads of competition policy and innovation: Comment on Ahlborn and Evans. *Antitrust Law Journal, 75*(3), 933–963.

Lee, R. S. T. (2020). Future trends in quantum finance. In *Quantum finance*. Springer. https://doi.org/10.1007/978-981-32-9796-8_14

Lem, S. (1964). *Bajki Robotów [Robot Fables]*. Wydawnictwo Literackie.

Lem, S. (1990). *The cyberiad: Fables for the cybernetic age, masterpieces of science fiction*. Easton Press, reprint edition.

Li, F., & Du, Y. (2018). From AlphaGo to power system AI: What engineers can learn from solving the most complex board game. *IEEE Power and Energy Magazine, 16*(2), 76–84. https://doi.org/10.1109/MPE.2017.2779554.

Loukis, E. N., Maragoudakis, M., & Kyriakou, N. (2020). Artificial intelligence-based public sector data analytics for economic crisis policymaking. *Transforming Government: People, Process and Policy, 14*(4), 639–662. https://doi.org/10.1108/TG-11-2019-0113

Lu, M., Chen, Q., Chen, Y., & Sun, W. (2020). Micromanagement in StarCraft Game AI: A case study. *Procedia Computer Science, 174*(2020), 518–523. https://doi.org/10.1016/j.procs.2020.06.119

Lytras, M. D., Sarirete, A., Visvizi, A., & Chui, K. W. (2021). *Artificial intelligence and big data analytics for smart healthcare*. Academic Press.

Lytras, M. D., Visvizi, A., Damiani, D., & Mthkour, H. (2018). The cognitive computing turn in education: Prospects and application. *Computers in Human Behavior*. https://doi.org/10.1016/j.chb.2018.11.011

Malik, R., Visvizi, A., & Skrzek-Lubasińska, M. (2021). The gig economy: Current issues, the debate, and the new avenues of research. *Sustainability, 13*, 5023. https://doi.org/10.3390/su13095023

Manne, G. A., & Wright, J. D. (2010). Innovation and the limits of antitrust. *Journal of Competition Law & Economics, 6*(1), 153–202. https://doi.org/10.1093/joclec/nhp032

Manners, I. (2002). Normative power Europe: A contradiction in terms? *JCMS: Journal of Common Market Studies, 40*, 235–258. https://doi.org/10.1111/1468-5965.00353.

Naudé, W. (2021). Artificial intelligence against COVID-19: An early review. IZA Discussion Paper, 13110. https://covid-19.iza.org/publications/dp13110/.

Neubert, M. J., & Montañez, G. D. (2020). Virtue as a framework for the design and use of artificial intelligence. *Business Horizons, 63*(2), 195–204. https://doi.org/10.1016/j.bushor.2019.11.001

Niebel, C. (2021). The impact of the general data protection regulation on innovation and the global political economy. *Computer Law & Security Review, 40*. https://doi.org/10.1016/j.clsr.2020.105523.

Nitzberg, M., & Zysman, J. (2021). Algorithms, data, and platforms: The diverse challenges of governing AI (March 10, 2021). *Journal of European Public Policy*. Available at SSRN https://ssrn.com/abstract=3802088 or https://doi.org/10.2139/ssrn.3802088.

OECD. (2015). *Frascati manual 2015: Guidelines for collecting and reporting data on research and experimental development, the measurement of scientific, technological and innovation activities.* OECD Publishing.https://doi.org/10.1787/9789264239012-en

Outeiral, C., Strahm, M., Shi, J., Morris, G. M., Benjamin, S. C., & Deane, C. M. (2021). The prospects of quantum computing in computational molecular biology. *WIREs Computational Molecular Science, 11*, e1481. https://doi.org/10.1002/wcms.1481

Piattini, M., Peterssen, G., & Pérez-Castillo, R. (2020). Quantum computing: A new software engineering golden age. *SIGSOFT Software Engineering Notes, 45*(3), 12–14. https://doi.org/10.1145/3402127.3402131

PWC. (2017). *Artificial intelligence study.* Price Waterhouse Coopers (PWC). https://www.pwc.com/gx/en/issues/data-and-analytics/publications/artificial-intelligence-study.html.

Rajoub, B. (2020). Supervised and unsupervised learning. In W. Zgallai (Ed.), *Developments in biomedical engineering and bioelectronics, biomedical signal processing and artificial intelligence in healthcare* (pp. 51–89). Academic Press. https://doi.org/10.1016/B978-0-12-818946-7.00003-2

Rakowski, R., Polak, P., & Kowalikova, P. (2021). Ethical aspects of the impact of AI: The status of humans in the Era of artificial intelligence. *Society.* https://doi.org/10.1007/s12115-021-00586-8

Roberts, H., Cowls, J., Morley, J., et al. (2021). The Chinese approach to artificial intelligence: An analysis of policy, ethics, and regulation. *AI & Society, 36*, 59–77. https://doi.org/10.1007/s00146-020-00992-2

Rosamond, B. (2003). Babylon and on? Globalization and international political economy. *Review of International Political Economy, 10*(4), 661–671. https://doi.org/10.1080/0969229031000160 1920

Ryan, M., & Stahl, B. C. (2021). Artificial intelligence ethics guidelines for developers and users: Clarifying their content and normative implications. *Journal of Information, Communication and Ethics in Society, 19*(1), 61–86. https://doi.org/10.1108/JICES-12-2019-0138

Sætra, H. S. (2020). Correction to: The parasitic nature of social AI: Sharing minds with the mindless. *Integrative Psychological & Behavioral Science, 54*(2), 327. https://doi.org/10.1007/s12124-020-09536-1

Sarirete, A., Balfagih, Z., Brahimi, T., Lytras, M. D., & Visvizi, A. (2021). Artificial intelligence and machine learning research: Towards digital transformation at a global scale. *Journal of Ambient Intelligence and Humanized Computing.* https://doi.org/10.1007/s12652-021-03168-y

Shu, C. (2014). *Google acquires artificial intelligence startup DeepMind for more than $500M.* Techcrunch, 27 January 2014. https://techcrunch.com/2014/01/26/google-deepmind/.

Sikos, L. F., & Choo, K. -K. R. (Eds.) (2020). *Data science in cybersecurity and cyberthreat intelligence.* Springer. https://www.springer.com/gp/book/9783030387877.

Smuha, N. A. (2021). From a 'race to AI' to a 'race to AI regulation': Regulatory competition for artificial intelligence. *Law, Innovation and Technology, 13*(1), 57–84. https://doi.org/10.1080/175 79961.2021.1898300

Strickland, E. (2019). How smart is artificial intelligence? *IEEE Spectrum*, 19 April 2019. https://ieeexplore.ieee.org/stamp/stamp.jsp?arnumber=8678419.

Russell, S., & Norvig, P. (2020). *Artificial intelligence: A modern approach* (3rd Ed.). Pearson. https://www.amazon.com/Artificial-Intelligence-Modern-Approach-3rd/dp/0136042597.

Troisi, O., Visvizi, A., & Grimaldi, M. (2021). The different shades of innovation emergence in smart service systems: The case of Italian cluster for aerospace technology. *Journal of Business & Industrial Marketing, ahead-of-print.* https://doi.org/10.1108/JBIM-02-2020-0091.

Turenne Slojander, C. (1996). The rhetoric of globalization: What's in a Wor(l)d? *International Journal, 51*(4), 603–616. Globalization, https://doi.org/10.2307/40203150.

Ulnicane, I., Knight, W., Leach, T., Carsten Stahl, B., & Wanjiku, W.-G. (2021). Framing governance for a contested emerging technology: Insights from AI policy. *Policy and Society, 40*(2), 158–177. https://doi.org/10.1080/14494035.2020.1855800

van Veenstra, A. F., Grommé, F., & Djafari, S. (2020). The use of public sector data analytics in the Netherlands. *Transforming Government: People, Process and Policy, ahead-of-print*(ahead-of-print). https://doi.org/10.1108/TG-09-2019-0095.

Vesnic-Alujevic, L., Nascimento, S., & Pólvora, A. (2020). Societal and ethical impacts of artificial intelligence: Critical notes on European policy frameworks. *Telecommunications Policy, 44*(6), 2020. https://doi.org/10.1016/j.telpol.2020.101961

Visvizi, A., & Bodziany, M. (Eds.). (2021). *Artificial intelligence and its context—Security.* Springer.

Visvizi, A., Daniela, L., & Chen, Ch. W. (2020). Beyond the ICT- and sustainability hypes: A case for quality education. *Computers in Human Behavior, 107.* https://doi.org/10.1016/j.chb.2020.106304.

Visvizi, A., Lytras, M. D., & Aljohani, N. (2021). Big data research for politics: Human centric big data research for policy making, politics, governance and democracy. *Journal of Ambient Intelligence and Humanized Computing, 12*(4), 4303–4304. https://doi.org/10.1007/s12652-021-03171-3

Xu, L. (2020). The Dilemma and countermeasures of AI in educational application. In *2020 4th International Conference on Computer Science and Artificial Intelligence* (pp. 289–294). Association for Computing Machinery. https://doi.org/10.1145/3445815.3445863.

Zhuang, Yt., Wu, F., Chen, C., & Pan, Y. (2017). Challenges and opportunities: From big data to knowledge in AI 2.0. *Frontiers of Information Technology & Electronic Engineering, 18*, 3–14. https://doi.org/10.1631/FITEE.1601883.

Anna Visvizi Ph.D., Associate Professor, SGH Warsaw School of Economics, and Visiting Scholar at Effat University, Jeddah. An economist and political scientist, editor, researcher and political consultant, Prof. Visvizi has extensive experience in academia, think-tank and government sectors in Europe and the US. Her expertise covers issues pertinent to the intersection of politics, economics and ICT. This translates in her research and advisory roles in the area of AI and geopolitics, smart cities and smart villages, knowledge & innovation management, and technology diffusion, especially with regards to the EU and BRI. ORCID: 0000-0003-3240-3771 Email: avisvi@sgh.waw.pl

A Typology of AI Applications in Politics

Henrik Skaug Sætra⊙

Abstract Politics relates to the most fundamental questions of humans' existence, and modern western societies tend to pride themselves on basing politics on democratic principles, including principles of inclusivity, participation, and autonomy and self-rule. Meanwhile, the continuously improving capabilities of various artificial intelligence (AI) based systems have led to the use of AI in an increasing amount of politically significant settings. AI is now also used in official government decision-making processes, and the political significance of using AI in the political system is the topic of this chapter. The purpose is to examine the potential for AI in official political settings, distinguishing between five types of AI used to *support*, *assist*, *alleviate*, *augment* or *supplant* decision makers. The chapter explores the practical and theoretical potential for AI applications in politics, and the main contribution of the chapter is the development of a typology that can guide the evaluation of the impacts of such applications. AI in politics has the potential to drastically change how politics is performed, and as politics entails answering some of society's most fundamental questions the article concludes that the issues here discussed should be subject to public processes, both in politics and society at large.

Keywords Artificial intelligence · Politics · Ethics · Technocracy · Typology

1 Introduction

Politics relates to the most fundamental questions of humans' existence, such as who should get what, how individuals can legitimately get anything, and who is allowed to do what when. Modern western societies tend to pride themselves on basing politics on democratic principles, including principles of inclusivity, participation, and autonomy and self-rule. However, modern democracies face a number of challenges, such as increased complexity purportedly requiring expert-rule (Dahl, 1985;

H. S. Sætra (✉)
Faculty of Business, Languages, and Social Sciences, Østfold University College, N-1757 Halden, Norway
e-mail: henrik.satra@hiof.no

© The Author(s), under exclusive license to Springer Nature Switzerland AG 2021
A. Visvizi and M. Bodziany (eds.), *Artificial Intelligence and Its Contexts*, Advanced Sciences and Technologies for Security Applications,
https://doi.org/10.1007/978-3-030-88972-2_3

Radaelli, 2017), polarisation (Sunstein, 2018), and a general decline in participation (Gray & Caul, 2000; Hooghe, 2014).

Meanwhile, the continuously improving capabilities of various artificial intelligence (AI) based systems have led to the use of AI in an increasing amount of politically significant settings. For example, machine learning algorithms based on deep personality profiles can be used to "nudge" individuals and allows for more effective influencing of individuals and their actions in general (Yeung, 2017). The combination of AI and big data can also be leveraged to influence political opinions and elections more directly (González, 2017). Such applications are politically significant, as they involve a sort of algorithmic governance that might entail political harms (Danaher et al., 2017; Sadowski & Selinger, 2014). However, AI is now also used in official government decision-making processes (de Sousa et al., 2019; Local Government Association, 2020; Veale & Brass, 2019). The political significance of using AI in the political system is the topic of this chapter, and not AI as a general phenomenon that has effects on the political system from the outside.

The purpose of this chapter is to examine the potential for AI in official political settings, distinguishing between five types of AI used to *support*, *assist*, *alleviate*, *augment* or *supplant* decision makers. When used to support, AI systems might for example serve the role of expert advisors that perform the role of analysis and preparation of scenario analyses. When used to supplant, however, AI is used in a way that can approach an AI technocracy (Sætra, 2020b). In between these extremes, there is a gradual movement from taking the machine into decision-making processes and granting it increasing amounts of autonomy. In terms of the typical "loop" metaphor, the five aforementioned types range from machine-out-of-the-loop to human-out-of-the-loop.

The chapter explores the practical and theoretical potential for AI applications in politics, and the main goal is to develop a typology that can guide the evaluation of the various positive and negative impacts that will have to be considered for such applications. Such an evaluation of the impacts of AI can in certain circumstances be performed in anticipation of the introduction of the AI systems. However, as AI is already used in politics the evaluation will in other instances entail analysing and judging systems already in place. The implications of the findings suggests that AI in politics has the potential to drastically change how politics is performed. As politics entails answering some of the most fundamental questions regarding right and wrong, and who gets what, the article concludes that the issues here discussed should be subject to public processes, both in politics and society at large. The argument is structured as follows. In Sect. 2, a brief examination of AI systems potential in political settings is presented. Section 3 contains the typology of AI in politics, where the typology is first introduced before the five types are presented in more detail. In this section, AI as a potential source of good is emphasised, in line with the UN's initiative "AI for good" (ITU, 2020). Section 4 explores the implications of the use of AI in politics more generally, as the main challenges discussed in relation to the five types are considered at a more general level with an emphasis on the need for social control of technology. Conclusions then follow.

2 AI and Politics

The reason AI in politics is becoming an increasingly relevant topic is that the capability of AI systems has dramatically improved along with advances in computing power and the availability of ever increasing data sets—big data (Marcus & Davis, 2019; Mayer-Schönberger & Cukier, 2013). While big data powered AI systems might not be able to surpass human beings in a range of creative and moral decision making processes (Sætra, 2018), they are exceedingly adept at solving the most complex and comprehensive problems related to strategy, calculation, and optimisation. Well-known example are playing the games of chess and go (Campbell et al., 2002; Chouard, 2016; Google, 2020). The linkages between such games and real-world problems of politics and war are ancient, and this chapter is based on the assumption that a range of political problems are the kind of problems AI excels at. Examples could be traffic management handling, optimising various forms of infrastructure, etc. Controlling a dynamic tax system is another proposed application of AI (Zheng et al., 2020). These are not hypothetical or far-fetched possibilities of AI use—AI is actually used in handling these and a wide range of other political problems already (de Sousa et al., 2019).

The technological aspects of AI used in political settings are not explored here. It suffices to note that the capabilities of AI here discussed are based on current or near-future technology, and not some speculative variety akin to the superintelligence, a singularity, etc. (Boström, 2014; Kurzweil, 2015). Current public sector AI applications are based on a variety of machine learning approaches, in which artificial neural networks are the most frequently used (de Sousa et al., 2019). Similar approaches are assumed to be the basis of the applications discussed in this chapter, and this is also the approach used in Google's AlphaZero software (Google, 2020).

While AI is already applied in political settings, there is much disagreement between the most optimistic and pessimistic commentators on AI's potential in politics (Katzenbach & Ulbricht, 2019). In addition to the question of ethical harms (discussed below), some argue that AI's potential is often overplayed, as purported benefits derived from experimental closed system settings will rarely translate into the promised effects when applied in the open and unpredictable systems of the real world (Sætra, 2021). However, AI is already employed and used in political contexts related to, amongst other areas, general public service, economic affairs, environmental protection, public health, transportation, education, policing and security, and the military (de Sousa et al., 2019). Predictive policing, immigration, and social benefits are other areas in which AI is already used (Janssen & Kuk, 2016; Katzenbach & Ulbricht, 2019). These examples suggest that while AI capabilities might at times be overplayed, there is also some merit in such claims.

2.1 AI Impact on Democracy

The goal of this chapter is not to recount the various use cases in which AI has been shown to be capable of performing a function related to political decision making, but rather to examine how the introduction of AI might impact our democratic systems. By leveraging new technologies in political decision making, we might see incremental changes to various seemingly separate parts of the political system, that together lead to more comprehensive changes to our political system. Micro level effects, such as the digitalisation of the tasks previously performed by various bureaucrats, or the automation of certain services in local administrative offices, might add up to macro level changes in which our democracy changes character.

Being aware of and analysing such changes are essential for making sure that societies control technology, and not the other way around. Small and incremental changes add up, but each of these changes rarely garner the academic or public attention required to gauge the public's opinions and willingness to accept the macro level changes that are associated with these changes. This relates to Collingridge's dilemma and the idea that new technologies are often easily controlled and regulated, but we do not really know the impacts of this technology at that stage. If we allow the technology to proliferate and grow, however, we will know the impacts, but we run the risk that the technology has been so widely integrated, and become integral to so many social functions, that it can now hardly be contained or controlled (Collingridge, 1980).

The questions involved in AI inroads into the political domain relate to fundamental questions of what democracy is, or should be. Questions of democracy are largely normative issues, such as whether democracy should be based on competition (Schumpeter, 2013) or deliberation (Elster, 1997; Gutmann & Thompson, 2009). The various forms of AI applications here discussed all have impacts related to such questions, and it will be shown that AI can be used is ways that foster either of these archetypes, as AI as a technology can be moulded and applied in a wide variety of ways.

A number of other challenges related to both cybersecurity and ethics are not considered in this chapter. However, AI ethics entails the analysis of a wide range of challenges and some of these are relevant when and if AI is applied in political settings. AI is most effective when combined with large and comprehensive data sets, and it is thus associated with a logic of surveillance and problems related to privacy (Sætra, 2019a, 2020a; Solove, 2000; Véliz, 2020). When AI and data is combined, improved means of influence and manipulation are created, and these are often related to the techniques of nudging (Sætra, 2019b; Yeung, 2017). Data is also associated with various types of bias and potential discrimination (Buolamwini & Gebru, 2018; Müller, 2020; Noble, 2018), and it is impossible to completely separate technology from power (Culpepper & Thelen, 2020; Gillespie, 2010; Sagers, 2019; Sattarov, 2019). Human beings are involved in all stages of systems and algorithm design and deployment, and in the process of turning human phenomena into data of various sorts (Janssen & Kuk, 2016; Ma, 2020; Sætra, 2018). This entails that data—and consequently AI—does not represent an objective or neutral arena of analysis and decision-making (Sætra, 2018).

3   Typology of AI Applications

This chapter presents a typology for categorising various AI applications in politics, as shown in Table 1. This typology emphasises the role of the "loop" often discussed in relation to human–machine interaction and the constellation of humans and machines in decision-making processes.

Several other typologies of AI in politics have been proposed. Katzenbach and Ulbricht (2019) created a typology for algorithmic governance, in which the dimensions of transparency (T) and level of automation (A) gave rise to four types of AI systems: autonomy-friendly systems (high T, low A), trust-based systems (low T, low A), licensed systems (high T, high A), and out-of-control systems (low T, high A). Another effort is that of Sadowski and Selinger (2014), whose taxonomy relates to technocracy, but in reality applies to the less comprehensive AI applications as well. Their taxonomy details different domains of application, and this chapter only details the political domain. They propose three potential means of AI influence: mandates, nudges, technical mediation. Most of the AI applications here discussed entail some form of technical mediation, but nudges are also relevant for most of the types, while mandates are only relevant for systems in which AI supplants human decision makers.

3.1   Support Democratic Decision-Making

The first type of AI application aims to support, or augment, human decision-makers (Veale & Brass, 2019). The most obvious form of AI support in decision making is the use of AI systems for analysis—often referred to as big data analytics (Chen et al., 2012). Big data, data mining, and machine learning are used to generate new insight in most sectors and areas of business (Dean, 2014). In a similar vein, AI can be used to improve analysis, optimisation, categorisations, and predictions in the domain of politics (Local Government Association, 2020). Economic policy would benefit from better analysis of the various patterns of income, investments, and profit generation, and systems that could predict and monitor the effects of different tax policies, for example, could improve policies (Zheng et al., 2020). When determining congestion tolls on roads, taxes on various products, the social benefits of banning unhealthy products, etc., AI has a clear potential for improving the foundations of decision making, and thus also decisions (Fig. 1).

Table 1 Typology for AI systems in political contexts. *Source* Author's own arrangement

Type	AI's role	Constellation of "decision loop"		Autonomy (A) and transparency (T)[1]	Means of governance[2]
Support	Prepare	Machine-out-of-the-loop		High T, low A	Technical mediation (M)
Assist	Automate and prepare	Human-in-the-loop		High/low T, high A	Technical mediation/Nudges
Alleviate	Automate	Human-on-the-loop		High/low T, high A	Technical mediation/Nudges
Augment	Guide and tutor	Machine-on-the-loop		Low T, Low, A	Technical mediation/Nudges
Supplant	Decide	Human-out-of-the-loop		Low T, high A	Mandates/nudges

[1] Based on the Katzenbach and Ulbricht (2019).
[2] Based on Sadowski and Selinger (2014).

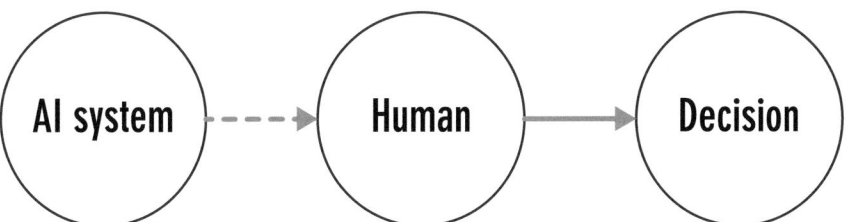

Fig. 1 Loop constellation for AI support systems. *Source* Author's own arrangement

Such AI systems would support human decision-makers. In the oft-used terminology of human in/on/out of-the-loop, such use of decision support does not necessarily require us to consider the *machine* to be in the loop at all. Human decision-makers are in full control, and AI analyses and predictions are merely some of many potential inputs. The main potential problems related to such applications of AI stem from a lack of transparency, challenges related to assigning responsibility, displacement of human expertise and consequently a limitation of human agency and involvement in political decisions, and the potential for biased and discriminatory results. These challenges are discussed in Sect. 4.

3.2 Assisting Humans

In addition to pure analysis and the preparation of materials that might factor into human decision-making processes, AI systems are increasingly being taken into the actual decision-making systems and processes (Veale & Brass, 2019). The purpose of such systems is not only to improve the foundations on which decisions are made, but also to help optimise and improve the processes themselves. This more comprehensive application of AI takes the machine into the loop, and involves machines in various stages of the formal decision-making processes. Rather than merely providing potential input for the decision makers, assistive AI involves integrating the machine in the loop by allowing machines to interact with the formal decisions-making process.

For example, a machine learning system might be applied in a country's social service system in order to analyse applicants for benefits. Following the training of the model, the system could proceed to place the applicants in different groups according to the predicted services required, and the machine might even prepare a proposed decision that is the forwarded to the bureaucrats in charge of formally determining if the applicant is to receive benefits or not. These sorts of applications are already widespread, and the Norwegian Labour and Welfare Administration has implemented systems akin to those here described (Ringnes, 2019) (Fig. 2).

These applications are typically referred to as human-in-the-loop, as human decision makers are integral and required in the decision-making process. The ethical

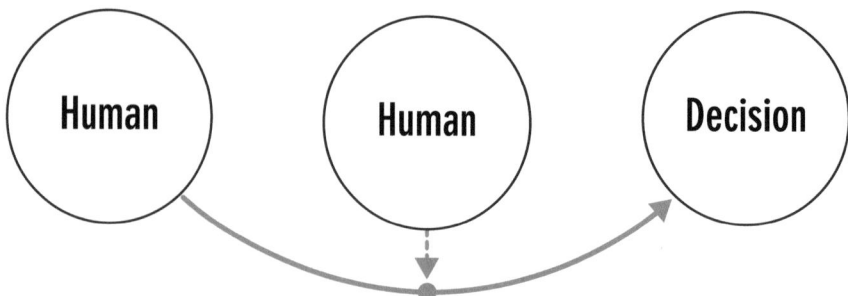

Fig. 2 Loop constellation for AI assistance systems. *Source* Author's own arrangement

challenges related to supporting systems applies here as well, but the challenges of transparency and displacement of human expertise are here exacerbated as humans might take on a mere rubber-stamping role that is not conducive to producing knowledge and understanding of the decisions made.

3.3 Alleviating Humans

While AI might be used to simply analyse or suggest policy, it can be applied in ways that fully automates certain decision-making processes (Veale & Brass, 2019). If we continue with the example of social benefits, an AI system that evaluates applications *and* makes decisions on its own would not simply assist human decision makers, but alleviate their decision-making loads. Such systems could in principle be applied to decisions concerning the determination of bail in criminal cases, the fines issued for traffic violations, the social benefits received, etc.

AI systems could also be used to make already automated systems—such as traffic lights, more dynamic and effective (Ma, 2020). In Norway, certain cities and zones also have dynamic speed limits, based on levels of pollution, congestion, etc., and AI systems might be used to fully automate such systems (Fig. 3).

Alleviation systems are distinguished from assistance systems in that the AI systems here makes decision without requiring human confirmation or control.

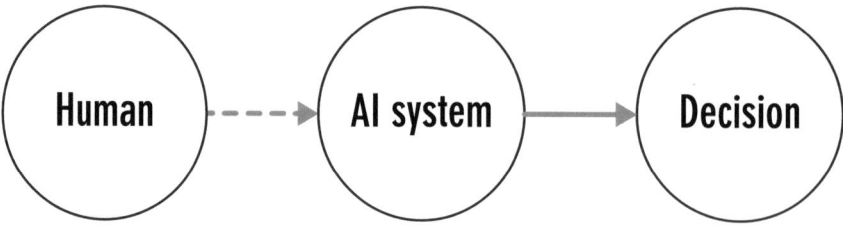

Fig. 3 Loop constellation for AI alleviation systems. *Source* Author's own arrangement

However, such systems are human-*on*-the-loop, as human decision makers are still responsible for the oversight of the system and the decisions AI systems make. The systems act on behalf of humans, so to speak, and humans are thus clearly still part of the loop, even though the human is no longer strictly necessary for decision-making. Compared to assistance systems, the ethical challenges are here further exacerbated, as increasing risks of bias and discrimination emanating from the AI systems become more and more opaque as the human decision makers is in practice distanced from the decision-making process. Furthermore, at this point AI systems begin to perform political functions to such a degree that issues of automation leading to job loss etc., becomes highly relevant.

3.4 Augmenting Human Decision Maskers

A different type of application of AI would be to develop and implement system in which AI performed the role of tutors or assistant for human political agents. While the other types mainly relate to practical bureaucratic decision-making, this sort of augmentation applies both to bureaucrats and all other political agents—citizens with a set of political rights. The description of this form of AI application is tentative, and further research is required to develop the potential for AI augmented political decision-making.

Through the use of learning analytics and AI, intelligent tutoring systems (ITS) could be developed to assist and aid all political agents in tasks related to various forms of voting or more active political decision-making (Erümit & Çetin, 2020; Humble & Mozelius, 2019). ITSs are built as responsive tools that learn the capabilities of a human learner and then choose the appropriate learning material and method of teaching for different situations. Such systems have proven their potential in education, and are likely to continue improving as both data sets and AI improves (Heaven, 2019; Nwana, 1990).

The key difference between this type and the others is that the AI system is aimed squarely at the human decision maker, and not the problem to be solved. AI will thus be implemented to help humans make better decisions, not through providing them with correct answers, but through improving the required skills and guiding them towards the required materials for understanding the problems. Policymakers working on economic policy would thus not use the AI economist of Zheng et al. (2020) to make policy. However, that AI system might be implemented as the expert module in an ITS geared towards teaching economics and policy (Fig. 4).

Using AI to augment human decision-makers has the potential of improving political decisions without some of the negative consequences associated with the aforementioned types of applications. Since AI is used to make humans autonomously reach better decisions, the decisions made, and policy implement, would be based on a rationale that some human being could in fact explain and defend. While the AI system might implement even better decisions had it been allowed to optimise fully, such decisions could be beyond the realm of human fathomability, and would thus

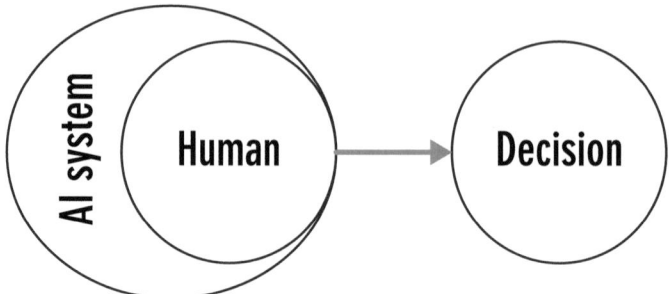

Fig. 4 Loop constellation for AI augmentation systems. *Source* Author's own arrangement

be problematic with regard to transparency. AI augmentation thus strikes a balance between improving decisions and maintaining transparency and meaningful human control of the decisions.

Such systems are premised on the idea that the AI system might learn to know the learner, and understand and anticipate their needs. They might involve the creation of something akin to "thought clones" that model the learner, and lets the AI system digitally experiment with the effects of various forms of educational approaches, etc. (Truby & Brown, 2020). Such systems might also be based on the creation of simulation systems for improving learner education, involving the creatin of "digital twins" (Bruynseels et al., 2018; Datta, 2017), a term coined to describe digital representations of physical systems, but now also used to describe digital representations of persons and living objects (Batty, 2018).

3.5 Supplant Democratic Decision-Making

Finally, there are the potential AI applications in which humans are taken out of the loop. While alleviating systems were also in principle fully automated, supplanting systems entail taking the human completely out of the decision-making loop and delegating authority to AI systems. However, human politicians will naturally still be in charge of setting the goals for the AI system, and determining the core values that the AI systems are to optimise (Janssen & Kuk, 2016; Sadowski & Selinger, 2014). They will, however, separate the political from the technical (Peters, 2014), and thus make humans responsible for political decision while AI systems are in charge of the technical implementation.

The AI systems applied in this manner resembles the technocrats of Meynaud (1969)—technically trained experts who "rule by virtue of their specialist knowledge and position in dominant political and economic institutions". This is distinct from both epistocracy (Estlund, 2003) and meritocracy (Arrow et al., 2018; Young, 1994), as technocrats are chosen because of skills in science and technology, and not on expertise in politics or particular mental or cognitive abilities (Sætra, 2020b).

While Schumpeter (2013) argued that government *by* the people would be prone to produce inferior political results, his solution was to replace "the people" with competent politicians who compete for power. It is possible to take his argument even further. As "the people" are prone to produce bad outcomes, politicians are people too, and we might achieve even better results if we remove people from the equation entirely. As people are prone to all sorts of self-interested or even egotistical behaviour, bias, cognitive limitations, and various other sources of error, some see machines as a potential way towards better decision-making (Kahneman, 2011; Mueller, 2003; Sætra, 2020b).

The main reason for choosing to implement such solutions would be to achieve some form of rational optimisation in politics. Just as computers now surpass humans as chess and go, and having humans in the loop when these systems play would simply be detrimental to the quality of play, this might also hold in parts of the political domain. Prime examples where such systems could be contemplated are those in which democratic politics is most inept at producing good solutions. Combatting climate change might be one such area, as the short-term nature of human politics is potentially overcome by the rational and strategic long-term perspective of an AI properly programmed (Sætra, 2020b). Other issues might be planning systems of roads and other infrastructures, for example, as these are issues in which many different stakeholders, in different political districts, are impacted in diverse ways, leading to problems of achieving what could be called "optimal" decisions form a societal perspective.

While such systems may seem far-fetched, the argument that we have moved quite a long way towards such applications of AI in a range of areas has been made (Sætra, 2020b). While current applications might be considered relatively trivial, various examples of how everything from tax policy to transportation and distributional issues might be automated, have been produced (Lee et al., 2019; Zheng et al., 2020) (Fig. 5).

Taking humans out of the loop entails highlighting and underscoring all the potential ethical challenges of using AI in politics. In particular, the very notion of democracy, participation, and self-rule is here challenged, as humans are taken out of the political decision-making loop. While democratic processes might indeed be messy and preclusive of what some experts might label optimal, this mess might indeed be the best solution for societies in which democracy is valued. Furthermore, some

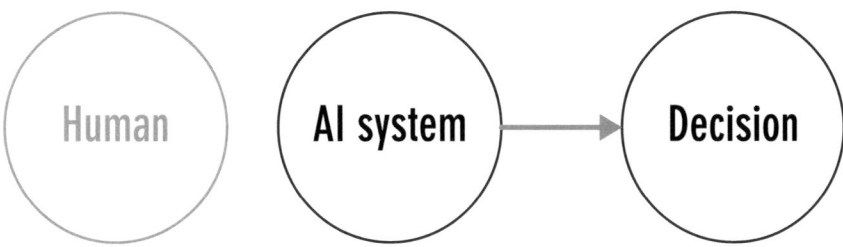

Fig. 5 Loop constellation for AI supplanting systems. *Source* Author's own arrangement

will argue that there are no such things as merely technical questions, or progress, and that all political decisions and processes are in reality value question in which society must strive for more popular control (Næss, 1989), not rational optimalisation and effectiveness. In addition, questions of redundancy systems, transparency, responsibility, and legitimacy are highly relevant for such systems (Sætra, 2020b).

4 AI in Politics: Contingencies and Caveats

As already introduced in the presentation of the five types, all forms of applications of AI in the political domain entail consequences of ethical significance. Considering the ethics of AI as it relates to politics should be a necessary first step whenever using AI in such ways are contemplated. The typology here presented allows for the systematic appraisal of such systems, and their associated ethical challenges.

The most direct challenges here discussed relate to the displacement of human expertise, attribution of responsibility for machine actions, the limitation of human areas of involvement and participation in politics, the decision of what are political and what are technical questions, the value and nature of democracy, political legitimacy, and the potential for machine-introduced or exacerbated bias and discrimination.

A detailed examination of these challenges are beyond the scope of these chapter, but they are discussed in detail in Sætra (2020b). However, there are also other effects of using AI in politics that are less often discussed, and these become the focus of the rest of this chapter. They relate to the potential subtle, indirect, and long-term effects of using AI.

Whenever technology is introduced in a new setting, the consequences are likely to have ripple effects beyond the intended and obvious micro level effects, even when such technological change is merely substitutional (Barley, 2020). As technology in introduced in a small area of a bureaucracy, opinions towards technology both within and outside the bureaucracies might change. In addition, when technology is introduced it might create certain dependencies, as humans who previously performed the tasks lose regular practice in performing the tasks taken over by machines. This might also affect skills in adjacent tasks and functions. As a consequence, the introduction of machines potentially creates the demand for more machines. Through such processes, digitalisation of the public sector might occur not through conscious and deliberate political processes and choices, but rather through *AI creep* (Sætra, 2020b).

Challenges related to privacy and surveillance, and in particular the collection creation of individual personality profiles, might be exacerbated if the public sector aims for AI applications that rely on such profiles for prediction and analysis. The creation of "thought clones" (Truby & Brown, 2020) is a riveting and fascinating idea—one that potentially opens the door for ultra-realistic and comprehensive digital experimentation is search for optimal policies. However, the potential use of such profiles for influencing and manipulating individuals is one reason for abstaining from such approaches (Sætra, 2019b; Yeung, 2017). An alternative approach would be to use agent-based modelling without the goal of accurately representing real individuals as agents. Agent-based modelling is a discipline in which much research has been conducted on the creation of artificial societies—including societies with a certain degree of politics (Boik, 2014; Epstein & Axtell, 1996). Using such simulations, as in the example of the AI economist (Zheng et al., 2020), alleviates some of the concerns related to privacy and surveillance.

Whenever a change affects societies' fundamental institutions, questions of how this affects different groups must be considered. For example, there are gender gaps in policy-making processes already (Funk & Gathmann, 2015), and digitalising such processes potentially influences and exacerbates such gaps. One concern is that the male-dominated world of technology produces systems and situations that in various ways are advantageous to men (Sattarov, 2019). This may occur to explicit or implicit bias in the design of the systems whenever the genders are not equally represented in the design and development of the systems. In addition to implicit or explicit bias in design, there is also the added risk that certain groups will be more inclined to or interested in taking advantage of the new opportunities provided by technology (Butler, 2000).

Facing these challenges in an appropriate manner is crucial for the development and application of AI in a sustainable (Sætra, 2021) and responsible (Dignum, 2019) manner. As the issues of AI in politics relate to the very structure of our political systems, and societies in general, citizens should be involved in whether and how AI is actually employed. One approach to stakeholder involvement in AI systems with societal—or *contested*—functions, is the idea of making systems with *society*-in-the loop (Rahwan, 2018).

Such proposals are in line with the idea proposed in this chapter, and the necessity of involving the public and society at large in decisions that may ultimately change the very nature of the political system. Based on this examination of the ethical implications of AI in politics, even incremental and seemingly small applications of various AI systems in politics are not innocuous or something that should be done without a broad and general debate the about the principles that should guide the use of AI in politics.

5   Conclusions

This chapter has presented a typology of AI applications in political settings, ranging from AI being used for simple analysis to AI becoming something akin to a technocrat. The goal of politics is not unidimensional, and this requires a thorough analysis of both the goals and preferences a society has for its political system and the effects of various AI applications. For example, a society might value both the efficiency of political decision-making and the final results produced while simultaneously valuing active and broad public participation and the preservation of basic democratic principles. These values might be compatible, but they need not be.

While AI can purportedly improve a range of politically relevant decision-making processes and decisions, such a position hinges on two controversial assumptions. First, the assumption that it is possible to label certain political questions as *technical* questions, and subsequently regard them as objects for optimisation without political relevance. This neglects how all decisions of a political nature will impact distribution of a wide variety of benefits and harms, and, as such, are inevitably questions of moral and political significance. Secondly, while isolated decisions might be improved by removing humans from the equation, such a move entails a loss of the positive benefits derived from having people actively involved in politics. While people might indeed make suboptimal decisions, what they learn, both regarding political processes and issues *and* each other through politics, is also potentially very valuable (Tocqueville, 2004).

The potential of AI in politics is great, and the fact that it is being implemented as rapidly as it is serves as proof of this. Nevertheless, the optimism regarding AI in politics must be tempered by the realisation that the potential negative impact is also quite significant. Making the required trade-off should be preceded by comprehensive and real stakeholder involvement, and this would in democratic societies entail broad public involvement, as societies first decide what their fundamental goals and values are (Næss, 1989), and only *then* decide whether or not using AI in politics is conducive to reaching these.

References

Arrow, K., Bowles, S., & Durlauf, S. N. (2018). *Meritocracy and economic inequality*: Princeton University Press.

Barley, S. R. (2020). *Work and technological change*. Oxford University Press.

Batty, M. (2018). Digital twins. *Environment and Planning B: Urban Analytics and City Science, 45*(5). https://doi.org/10.1177/2399808318796416.

Boik, J. C. (2014). *Economic direct democracy: A framework to end poverty and maximize well-being*. Book.

Boström, N. (2014). *Superintelligence: Paths, dangers, strategies*. Oxford University Press.

Bruynseels, K., Santoni de Sio, F., & van den Hoven, J. (2018). Digital twins in health care: Ethical implications of an emerging engineering paradigm. *Frontiers in Genetics, 9*, 31.

Buolamwini, J., & Gebru, T. (2018). Gender shades: Intersectional accuracy disparities in commercial gender classification. *Paper presented at the Conference on fairness, accountability and transparency*.

Butler, D. (2000). Gender, girls, and computer technology: What's the status now? *The Clearing House, 73*(4), 225–229.

Campbell, M., Hoane, A. J., Jr., & Hsu, F.-H. (2002). Deep blue. *Artificial Intelligence, 134*(1–2), 57–83.

Chen, H., Chiang, R. H., & Storey, V. C. (2012). Business intelligence and analytics: From big data to big impact. *MIS Quarterly*, 1165–1188.https://doi.org/10.2307/41703503.

Chouard, T. (2016). *The Go files: AI computer wraps up 4–1 victory against human champion*. Nature News.

Collingridge, D. (1980). *The social control of technology*. Frances Pinter.

Culpepper, P. D., & Thelen, K. (2020). Are we all amazon primed? Consumers and the politics of platform power. *Comparative Political Studies, 53*(2), 288–318.

Dahl, R. A. (1985). *Controlling nuclear weapons: Democracy versus guardianship.* Syracuse University Press.

Danaher, J., Hogan, M. J., Noone, C., Kennedy, R., Behan, A., De Paor, A., Morison, J., et al. (2017). Algorithmic governance: Developing a research agenda through the power of collective intelligence. *Big Data & Society, 4*(2), 2053951717726554.

Datta, S. (2017). Emergence of digital twins. *Journal of Innovation Management, 5*, 14–34.

de Sousa, W. G., de Melo, E. R. P., Bermejo, P. H. D. S., Farias, R. A. S., & Gomes, A. O. (2019). How and where is artificial intelligence in the public sector going? A literature review and research agenda. *Government Information Quarterly, 36*(4), 101392.

Dean, J. (2014). *Big data, data mining, and machine learning: Value creation for business leaders and practitioners.* Wiley.

Dignum, V. (2019). *Responsible artificial intelligence: How to develop and use AI in a responsible way.* Springer.

Elster, J. (1997). The market and the forum: three varieties of political theory. In *Deliberative democracy: Essays on reason and politics* (Vol. 3, p. 18).

Epstein, J. M., & Axtell, R. (1996). *Growing artificial societies: social science from the bottom up.* Brookings Institution Press.

Erümit, A. K., & Çetin, İ. (2020). Design framework of adaptive intelligent tutoring systems. *Education and Information Technologies*, 1–24.

Estlund, D. (2003). Why not epistocracy? In *Desire, identity and existence: Essays in honor of TM Penner* (pp. 53–69).

Funk, P., & Gathmann, C. (2015). Gender gaps in policy making: Evidence from direct democracy in Switzerland. *Economic Policy, 30*(81), 141–181.

Gillespie, T. (2010). The politics of 'platforms.' *New Media & Society, 12*(3), 347–364.

González, R. J. (2017). Hacking the citizenry? Personality profiling, 'big data' and the election of Donald Trump. *Anthropology Today, 33*(3), 9–12.

Google. (2020). AlphaZero: Shedding new light on the grand games of chess, shogi and Go. Retrieved from https://deepmind.com/blog/article/alphazero-shedding-new-light-grand-games-chess-shogi-and-go.

Gray, M., & Caul, M. (2000). Declining voter turnout in advanced industrial democracies, 1950 to 1997: The effects of declining group mobilization. *Comparative Political Studies, 33*(9), 1091–1122.

Gutmann, A., & Thompson, D. F. (2009). *Why deliberative democracy?* Princeton University Press.

Heaven, D. (2019). Two minds are better than one. *New Scientist, 243*(3244), 38–41.

Hooghe, M. (2014). Citizenship and participation. In L. De Luc, R. Niemi, & P. Norris (Eds.), *Comparing democracies.* Sage.

Humble, N., & Mozelius, P. (2019). *Teacher-supported AI or AI-supported teachers?* Paper presented at the European Conference on the Impact of Artificial Intelligence and Robotics 2019 (ECIAIR 2019), Oxford, UK.

ITU. (2020). AI4Good Global Summit. Retrieved from https://aiforgood.itu.int

Janssen, M., & Kuk, G. (2016). The challenges and limits of big data algorithms in technocratic governance. Elsevier.

Kahneman, D. (2011). *Thinking, fast and slow.* Macmillan.

Katzenbach, C., & Ulbricht, L. (2019). Algorithmic governance. *Internet Policy Review, 8*(4), 1–18.

Kurzweil, R. (2015). Superintelligence and Singularity. In S. Schneider (Ed.), *Science fiction and philosophy: From time travel to superintelligence* (pp. 146–170). Wiley-Blackwell.

Lee, M. K., Kusbit, D., Kahng, A., Kim, J. T., Yuan, X., Chan, A., …, Psomas, A. (2019). WeBuildAI: Participatory framework for algorithmic governance. In *Proceedings of the ACM on Human-Computer Interaction* (Vol. *3*(CSCW), pp. 1–35).

Local Government Association. (2020). Using predictive analytics in local public services. *Research.* Retrieved from https://www.local.gov.uk/using-predictive-analytics-local-public-services

Ma, M. (2020). The Law's new language? *Harvard International Law Journal Frontiers, 61.* Retrieved from https://harvardilj.org/2020/04/the-laws-new-language/

Marcus, G., & Davis, E. (2019). *Rebooting AI: building artificial intelligence we can trust.* Pantheon.

Mayer-Schönberger, V., & Cukier, K. (2013). *Big data: A revolution that will transform how we live, work, and think.* Houghton Mifflin Harcourt.

Meynaud, J. (1969). *Technocracy.* Free Press.

Mueller, D. C. (2003). *Public choice III.* Cambridge University Press.

Müller, V. C. (2020). Ethics of artificial intelligence and robotics. In E. N. Zalta (Ed.), *Stanford encyclopedia of philosophy* (Summer 2020 Ed.).

Næss, A. (1989). *Ecology, community and lifestyle: Outline of an ecosophy.* Cambridge university press.

Noble, S. U. (2018). *Algorithms of oppression: How search engines reinforce racism.* University Press.

Nwana, H. S. (1990). Intelligent tutoring systems: An overview. *Artificial Intelligence Review, 4*(4), 251–277.

Peters, B. G. (2014). *Politics of bureaucracy.* Routledge.

Radaelli, C. M. (2017). *Technocracy in the European Union.* Routledge.

Rahwan, I. (2018). Society-in-the-loop: Programming the algorithmic social contract. *Ethics and Information Technology, 20*(1), 5–14.

Ringnes, I. F. (2019, March 18th). *Kunstig intelligens kan hjelpe NAV med å gi bedre tjenester.* Mennesker og muligheter.

Sadowski, J., & Selinger, E. (2014). Creating a taxonomic tool for technocracy and applying it to Silicon Valley. *Technology in Society, 38,* 161–168.

Sætra, H. S. (2018). Science as a vocation in the era of big data: The philosophy of science behind big data and humanity's continued part in science. *Integrative Psychological and Behavioral Science, 52*(4), 508–522.

Sætra, H. S. (2019a). Freedom under the gaze of Big Brother: Preparing the grounds for a liberal defence of privacy in the era of Big Data. *Technology in Society, 58,* 101160.

Sætra, H. S. (2019b). When nudge comes to shove: Liberty and nudging in the era of big data. *Technology in Society, 59,* 101130.

Sætra, H. S. (2020). Privacy as an aggregate public good. *Technology in Society, 63,* 101422. https://doi.org/10.1016/j.techsoc.2020.101422

Sætra, H. S. (2020b). A shallow defence of a technocracy of artificial intelligence: Examining the political harms of algorithmic governance in the domain of government. *Technology in Society,* 101283.

Sætra, H. S. (2021). AI in context and the sustainable development goals: Factoring in the unsustainability of the sociotechnical system. *Sustainability, 13*(4). https://doi.org/10.3390/su13041738

Sagers, C. (2019). Antitrust and tech monopoly: A general introduction to competition problems in big data platforms: Testimony before the committee on the Judiciary of the Ohio Senate. Available at SSRN 3471823.

Sattarov, F. (2019). *Power and technology: A philosophical and ethical analysis.* Rowman & Littlefield.

Schumpeter, J. A. (2013). *Capitalism, socialism and democracy.* Routledge.

Solove, D. J. (2000). Privacy and power: Computer databases and metaphors for information privacy. *Stanford Law Review, 53,* 1393.

Sunstein, C. R. (2018). *# Republic: Divided democracy in the age of social media.* Princeton University Press.

Tocqueville, A. D. (2004). *Democracy in America.* The Library of America.

Truby, J., & Brown, R. (2020). Human digital thought clones: The Holy Grail of artificial intelligence for big data. *Information & Communications Technology Law,* 1–29.

Veale, M., & Brass, I. (2019). Administration by algorithm? Public management meets public sector machine learning. In K. Yeung & M. Lodge (Eds.), *Algorithmic regulation* (pp. 121–149). Oxford University Press.

Véliz, C. (2020). *Privacy is power*. Bantam Press.

Yeung, K. (2017). 'Hypernudge': Big Data as a mode of regulation by design. *Information, Communication & Society, 20*(1), 118–136.

Young, M. D. (1994). *The rise of the meritocracy*. Transaction Publishers.

Zheng, S., Trott, A., Srinivasa, S., Naik, N., Gruesbeck, M., Parkes, D. C., & Socher, R. (2020). The ai economist: Improving equality and productivity with ai-driven tax policies. *arXiv preprint* arXiv:2004.13332.

Henrik Skaug Sætra Ph.D. in Political Science, Associate Professor, Senior Manager, Sustainability, KPMG Norway. Research areas: ethics of artificial intelligence and big data, sustainability, environmental ethics, social robots, political theory, strategy and game theory. Email: henrik. satra@hiof.no ORCID: 0000-0002-7558-6451

Ethical Conditions of the Use of Artificial Intelligence in the Modern Battlefield—Towards the "Modern Culture of Killing"

Marek Bodziany ⓘ

Abstract The use of artificial intelligence (AI) on the modern battlefield causes justi-fied semantic and ethical controversies. Semantic, i.e., relating to meaning, contro-versies are related to delineating artificial intelligence (AI) used by a human during military operations, e.g., precision weapons capable of independent decision-making about the choice of target; and completely autonomous machines, i.e. thinking and deciding autonomously/independently from the human being. In this broad, and still nascent filed, several legal and ethical challenges emerge. This chapter sheds light on the emerging ethical considerations pertinent to the use of robots and other machines relying on unsupervised learning in resolving conflicts and crises of strategic importance for security.

Keywords Robots · Artificial Intelligence (AI) · Unsupervised learning · "the death culture" · Ethics and AI · "the ethics of killing"

1 Introduction

The ethics of warfare with the use of artificial intelligence (AI) can be considered at two levels. The first—technological—is associated with the human's endeavor to achieve a higher development level in all areas of life. One of them is the military sphere, including the development of military technologies necessary to provide the state not only with guarantees of domestic security but also with "national power" or, in a broader context, with state power (cf. Mearsheimer, 2001; Morgenthau, 1973). The second level concerns the axiological sphere in which "applied ethics" locates, focusing on the practical aspects of applying AI in everyday life (European Commission, 2019, p. 37, Madiega, 2019).

The problem concerns the "independence" of AI-enhanced, intelligent, systems, the so-called "narrow," of robots designed to perform a specific task, but also dilemmas related to artificial intelligence, the so-called "general," "real," or "strong"

M. Bodziany (✉)
Military University of Land Forces, Wrocław, Poland
e-mail: m.bodziany@interia.pl

AGI (Artificial General Intelligence). AGI refers to the claim that it is possible to create machines that perform tasks so skillfully that the underlying it thinking process is comparable to that of the human brain (Bird et al., 2020). Hence, devices have the ability to create patterns that can be used to solve problems, etc. (Różanowski, 2007, pp. 3–4; Goertzel & Pennachin, 2007), which can be compared to human intelligence. Artificial Intelligence Ethics deals with recommendations for manufacturers in the field of methods for minimizing moral damage resulting from bad (unethical) design and inappropriate application or use of AI (Bird et al., 2020).

Over the past years, driven by academic research on ethics in AI, as well as by the inroads of AI-enhanced technologies in everyday life, an increased interest in AI has been observed. The spread of AI revealed a whole range of new threats and opportunities to society. This forced governments to make first steps toward building regulatory frameworks addressing the ethical principles of AI research, AI development and AI use. The importance of the matter is evidenced by the fact that since January 2017, at least 22 different sets of ethical principles have been published (Bird et al., 2020) which focus on, among other things, social and economic problems. From this emerges the problem of replacing people by AI-based solutions in the labor market (cf. Frey & Osborne, 2013; Malik et al., 2021; Troisi et al., 2021), but also impoverishment of society, growing inequalities and social distance, as well as prejudices, discrimination and negative attitudes (Leslie, 2019, p. 4). Another important ethical issue is human rights and human dignity, as well as the identification of the boundary between 'good' and 'evil' that the use of AI entails.

The ethical challenge of using AI is revealed in its relations with humans, especially in the case of the replication of human behavior by a robot. Referred to in the literature as anthropoformism (from Greek: άνθρωπος + φόρμα) equated with falsification and / or identity theft (Remmers, 2019, p. 62; Darling 2016; Coeckelbergh, 2011). In the field of psychology, research on AI focuses on the influence of AI on the *psyche* and human behavior, which, being very unpredictable, becomes a massive challenge for engineering dealing with their replication (Bostrom & Yudkowsky, 2014, p. 319).

The issues raised constitute a prelude to the considerations on the ethics of using AI in conflicts and wars, which is often identified only with robotization, i.e., "narrow" artificial intelligence. In this context, the objective of this chapter is to examine the effects of using AI in wars, especially as viewed through the lens of ethical implications of the AI-use in modern warfare. Therefore, there were several questions concern the directions of development of "warlike" AI, (e.g., super artificial Intelligence), limitations in replacing humans with robots in the decision loop, ethical effects of using AI on the contemporary battlefield, as well as the role and tasks of international institutions in limiting AI development and setting clear legal and ethical boundaries.

Research problems allowed to identify the main hypothesis assuming that the development of AI currently has no clear limits to development and its effects in wars are of a dual nature. On the one hand, it poses a threat due to the occurrence of massive physical losses and the disappearance of human liability for damages caused by robots. On the other hand, it has a non-destructive dimension, which is concerned with the precise elimination of targets without suffering collateral damage (Boussios & Visvizi, 2017; Netczuk-Gwoździewicz, 2020; Pathe Duarte, 2020). Whether, in the face of the AI application on the modern battlefield, we are dealing with a "modern culture of killing" that is similar to computer games, where the goal—human becomes objectified and at the same time virtual creature? Although the answer to this question is complex and contentious, arguments confirming its validity will be presented later in this chapter. The issues included in the hypotheses are an introduction to the analysis presented later in the article. The first chapter develops the previously raised key concepts and approaches about AI contained in Polish and foreign literature. The second chapter is a discussion on the "robolution" and the directions of its development. The third chapter deals with the axiological implications of using artificial intelligence in war. All considerations end with conclusions.

2 Ethics of "War" Artificial Intelligence—Theoretical Implications and Source Overview

Contemporary considerations on AI the battlefield are related to three trends (Wong, Yurchak, Button et al., 2020). The first trend is the result of a change in approach to conflicts and wars after September 11, 2001, especially in Iraq and Afghanistan. Both conflicts radically changed the image of the battlefield of that time, from classical and symmetric to asymmetric and network-centric, based on distributed and coordinated advanced technologies. It was then favored by the development of precision weapons and unmanned aerial reconnaissance and destruction systems—commonly known as drones, and the dissemination and use of ground-based robots to combat improvised explosives. The first trend is characterized by developing human-driven robots (of various types and purposes) without their capability of autonomous decision-making. The second trend is related to the development of artificial intelligence (AI) in areas such as battlefield visualization, planning, machine learning, natural language processing, and robotics. The technology of autonomous vehicles implemented by the Defense Advanced Research Projects Agency (DARPA) has also developed significantly. The third trend concerns the increase in expenditure on war robotization and the development of "war" artificial intelligence in major players' strategies in world politics (Grosz et al., 2015).

The reasons for the development presented above constitute a prelude to considering the ethical aspects of artificial intelligence in the contemporary battlefield. Among the myriad definitions of AI, the popular description refers to it as the 'machines' (or computers) ability of independent thinking and/or problem-solving capacities. An important issue in the analysis of ethical areas of AI is the ontology of the sub-function of intelligent behavior or problem solving by robots or "intelligent" machines. Belong to them:

– searching for solutions to problems in a given space;

- pattern recognition based on developing internal representations of relevant solution patterns;
- generalization-based learning from past solutions to inform about future decisions;
- planning: organizing attention or resources to achieve a specific goal;
- induction based on adaptation and transfer of learned behaviors to a variety of environments. (Minsky, 1961, pp. 8–10; Wong, Yurchak, Button et al., 2020).

The afore-mentioned features, although not directly, constitute a premise for a public debate on the legitimacy of "endowing" machines with features, attributes, and abilities specific to human beings so that they have feelings, sensitivity, imagination, show empathy and observe not only the law but also the principles of social coexistence. The argument raised is correct to a large extent only in theory as the history of wars and the number of their victims prove that in favorable circumstances, a human shows a greater tendency to killing ($\theta\acute{\alpha}\nu\alpha\tau o\varsigma$) than feelings, sensitivity, and respect for other people ($\varepsilon\rho\omega\varsigma$). Human weakness on this level prompts reflection on the sense of replacing them with an intelligent machine programmed to act in the highest ethical, humanitarian, and precise way, even if it is the death of the enemy. Here is an ethical dilemma to which we will probably never see a solution. It is no less justified to analyze the approaches to the ethics of using AI in contemporary conflicts and wars and those that will take place in the future.

With regard to the conceptualization of AI in the decision-making process on the battlefield is the philosophy of a man entangled in a decision loop in relation to AI. It manifests itself in three forms. The first concerns the human "in the loop" (HITL), the essence of which is people's cooperation with artificial intelligence in the decision-making process, especially in the case of deciding about the use of lethal means of combat. The second one, called "on the loop" (HOTL), concerns routine AI decision-making, but humans monitor it, even in situations that determine people's lives. The third concept—the philosophy of "outside the loop" of people (HOOTL)—is based on independent decision-making without the participation or consent of people to the so-called "own" responsibility of the machine/system (Wong, Yurchak, Button, 2020).

To understand the ethical pillars of AI use it is useful to refer to the concept of "post-heroic mentality" (Luttwak, 1995). In the context of the Western culture of war, the concept highlights the disappearance of the warrior ethos and the natural tendency to avoid the risk of death as the result of warfare (Luttwak, 1995). Modern "war" technologies, including long-range weapons with a high level of precision or robotization of war, reveal this tendency, aiming at the maximum effectiveness of the technologies used with own minimal losses. In simplified terms, it can be assumed that this very premise has become the "flywheel" for the AI development for the needs of contemporary armed conflicts and building state power based on technological advantage.

Countless scientific texts have been written on artificial intelligence understood as "war" technology, and many reports and national and transnational programs of its development, design, purpose, and use in contemporary armed conflicts have been prepared. The document entitled "Artificial Intelligence and the Future of Defense" exemplifies the AI development into six phases. The first one, dated 1956–75, is named "Early Enthusiasm": The First AI Spring. The second covers the period 1974–80 and is called "Grandiose Objectives": The First AI Winter. The third phase, "Expert Systems": The Second AI Spring, spanned the years 1980–87. The fourth one from 1987 to 93 was called "Not the Next Wave,"; The Second AI Winter. The fifth, called "A Third, Sustained AI Spring, fell in the years 1993–2011, and the sixth, until now, is called "Big Data, Deep Learning, and an Artificial Intelligence Revolution" (De Spiegeleire, Maas, Sweijs, 2017, pp. 31–39; Safder et al., 2020). With AI development, the interest in ethical problems of its use in conflicts and wars has increased in the last two phases. Particular controversies have arisen from the interest in neural networks and genetic algorithms and the issues of AI-equipped "machines" independent thinking and decision-making. The last phase opened a new chapter in research on the use of artificial intelligence in conflicts and wars. It was possible thanks to advances in neurobiology and computer science and new techniques of incorporating neurons into network pattern recognition, as well as significant financial investments in artificial intelligence research, among others, by the Agency for Advanced Defense Research Projects—DARPA (Hinton & Salakhutdinov, 2006, pp. 504–507, Kamieński, 2014).

The interest in the issues of the ethics of employing AI in wars was also revealed among theologians and clergy, who exemplified the ethical areas of using AI with an emphasis on positive, adverse, neutral, mixed and/or ambiguous, and sometimes dual effects. In addition to AI appropriating a wide variety of everyday problems, theologians express concern about the use of drones, robots, and other "thinking machines" in wars, as well as problems in neurobiology and genetics (Green, 2016, pp. 10–11). It should be clearly emphasized that the ethics of using AI is not only a subject of theologians' and clergy' debates. It has also received many valuable publications primarily in the field of military use. It should be clearly emphasized that the ethics of applying artificial intelligence is not only a subject of debates by theologians and clergy. It found its place in the strategic documents of the US army, in which the rules of using AI were published, the threats associated with it were identified, as well as the directions of development (De Spiegeleire, Maas, Sweijs, 2017; Cummings, 2017).

The ethical dimension is associated with the issues of dehumanization of war and the blurred role of humans in the decision loop in relations with the robot (Kopeć, 2016), as well as with a whole range of legal provisions and recommendations in the case of using AI (Lucas, 2014, pp. 317–339; Remmers, 2019). Its essence is a philosophical reflection on the similarities in the behavior of humans and robots, and the broadly understood issues of anthropomorphism in human–machine relations. Very important are the matters of control over the deadly effects of the use of AI and robotics in war and not only in it (Arkin, 2011). The author presents recommendations for reducing unethical robotic behavior, ethical AI behavioral design, the use of affective functions as an adaptive component in the event of unethical behavior, and other issues related to minimizing the damage caused by robots.

Important for the development of the ethics of using AI in wars is the United States' (US) strategy for the development of AI, which is the result of the state's desire to maintain an uninterrupted technological advantage over the rest of the world while maintaining the five ethical principles of using AI: responsibility, equity, traceability, reliability and manageability (DoD, 2019). AI has also found its place in RAND Corporation's research on the importance of the US nuclear program (Boulanin, 2019; Geist & Andrew, 2018). Indeed, several authors express their legitimate concerns about military strategists who uncritically believe in the omnipotence and infallibility of AI (cf. Visvizi & Lytras, 2019; Visvizi, Lytras, Daniela, 2020).

3   Directions of "Robolution"

Considerations on the robotization of war and its axiological implications are inextricably linked with the contractual notion known in the literature as "robolution" (Windeck, 2014, p. 5). This concept is a semantic hybrid made of two words: robotization and revolution. Its existence is confirmed by the rapid development of robotics in the twenty-first century, which gained importance in 2008 when about 5.5 million working robots were recorded in various fields of the economy. The then forecasts predicted an increase in their number in 2011 to the level of 11.5 million (IFR, 2008), by 2016 by another 1.8 million, and by 2020 by another 3 million. Other data confirming the "robolution" show that as of 2016, there were five major markets worldwide accounting for 74% of global sales—China, South Korea, Japan, the United States, and Germany (IFR, 2020), while 21% of all robots are produced and employed in Asia. The arrangement proves the technological asymmetry of a group of hegemony-oriented countries in creating a new world order, whose ethical pillars will depend on the level of involvement of artificial intelligence in social life, economy, politics, media, and, above all, in the military sphere.

"Robolution" has changed not only the attitudes towards warfare but revealed its new "actor" that carries out orders and decides about the choice of the course of action (Danet & Hanon, 2014, p. XIII; Kopeć, 2016, p. 37). This entity is nothing more than the so-called "narrow" robot equipped with artificial intelligence, which is the equivalent of a soldier on the modern battlefield. The military "robolution" is revealed by the scale of involvement of land and mobile air robots in the conflicts in Iraq and Afghanistan. Their number exceeded 5,000 of various types and purposes (Sharkey, 2008). Currently, robotization of war is one of the critical directions of research focused on creating intelligent machines—humanoid robots capable of imitating human behavior and using them optimally and most effectively in warfare.

The largest expenditure on AI development is currently borne by the US Department of Defense, which spent $ 7.4 billion on its development in 2017 (Wong, Yurchak, Button, 2020). A similar trend is also taking place in China, which announced the "Next Generational Artificial Intelligence Development Plan" in 2017. Its essence was to achieve supremacy in AI by 2020, which confirms the value of this industry estimated at over $ 21.7 billion and taking over the strategic initiative by 2030 (Sayler, 2020, pp. 20–21; Özdemir, 2019, pp. 18–20). The third power interested in developing "war" artificial intelligence is Russia, which has created many organizations dealing with the development of artificial military intelligence for several years. In March 2018, it launched an artificial intelligence program, the primary development vector of creating the "AI and Big Data" consortium, a dedicated research laboratory, and the National Center for Artificial Intelligence (Bendett, 2018). Besides, Russia has created a defense research organization equivalent to the DARPA agency for autonomy and robotics, called the Foundation for Advanced Studies. It also hosted the annual conference on "Robotization of the Armed Forces of the Russian Federation" (Bendett, 2017). The Russian Armed Forces also research many applications of artificial intelligence, with particular emphasis on semi-autonomous and autonomous vehicles and information warfare and propaganda (Sayler, 2020, p. 24). The AI research programs conducted by Iran, Israel, Japan, and South Korea are also important.

The robotization of war is gaining more importance in two spheres. The first one is strictly combat technology, including intelligent precision weapons designed to destroy distant targets. The second one concerns intelligent humanoid robots intended for contact with the enemy or in challenging conditions. The second sphere concerns "machines" equipped with human-like intelligence and skills beyond the soldier's capabilities. Due to the limitation of the article, the argument on precision weapons will be omitted in favor of autonomous land vehicles and humanoid robots, the most recognizable representative of which is the ATLAS robot designed and built as part of the DARPA competition and currently produced by the Boston Dynamics concern (Antal, 2016). It is a product that is timeless and, at the same time, opens a new chapter in the development of robots and can move without power cables, maintain balance, lift weights, run, jump and even perform a full somersault and get up from a lying position. The most innovative designs are more advanced because, in addition to recognizing the enemy and reading their intentions, they have pre-programmed abilities of several different fighting styles and are equipped with many types of weapons. Although these robots are not widely used, they mark the azimuth of the armed forces' development towards cyborgization of war, similar to those found in fantasy films. They are associated with another question about the cyborgization of the future war and its potential social consequences.

Currently, mobile robots—autonomous vehicles designed to perform combat tasks independently or with remote control systems—are more common. One of them is the MAARS (Modular Advanced Armed Robotic System)—an autonomous combat module, which has excellent firepower and maneuvering capabilities liked human-crewed vehicles. Its modular design allows the robot to be equipped with various weapons, from blinding lasers to tear gas and even a grenade launcher. MAARS is a development of the SWORDS project that served in Iraq a few years ago. Another robot of this type is the GLADIATOR tactical unmanned armored ground vehicle, designed to support the US Marine Corps in various operations. A significant direction of research on the robotization of war is the miniaturization of robots, represented by a conceptual invention of a robot called the RoboBee X-Wing Tiny Flying Insect Robot developed at Harvard Microbiotics Laboratory. It is a small flying robot designed for surveillance and eavesdropping. Although the prototype does not do very well without a power cord, its design is an essential step towards insect-like robots capable of transmitting audio or video for intelligence purposes (Antal, 2016).

In other armed forces, there has also been significant progress in robotization and the use of AI on the battlefield. The Russian weapon manufacturer, KB Integrated Systems (KBIS), has developed the RS1A3 Mini Rex remote-controlled system—a lightweight tactical combat robot that fits the size of a soldier's backpack. Another Russian project is the Taifun-M autonomous reconnaissance platform designed to protect missile bases. A more advanced AI-equipped combat platform is the Uran-9 UGCV tracked vehicle. It has a very versatile armament: a 7.62 mm rifle, a 30 mm 2A72 cannon, an M120 "Attack" anti-tank guided missile or an anti-aircraft "Needle" (Russian игла) or "Arrow" (Russian стрела). It is a human-driven system with limited decision-making possibilities. Chinese robot designs refer to American ideas. The analogies are especially visible in the Sharp Claw and Sharp Claw 1 autonomous vehicles, which resemble the American MAARS system. A completely different concept is the Claw 2 system, whose purpose, size, and capabilities indicate an entirely new Chinese arms industry product. The interest in robots in the Iranian and Israeli Armed Forces is also very noticeable (Antal, 2016). Progress in this area is also visible in the Polish Armed Forces, which are equipped with, among others, "Balsa" type robots supporting the sappers' work in the removal of charges and dangerous materials, and "Inspector" and "Expert" reconnaissance and inspection robots used to neutralize explosives. In addition to AI-equipped land autonomous vehicles, most modern armed forces have implemented or are in the process of deploying intelligent flying platforms (autonomous or semi-autonomous) commonly known as drones. In this area, the armed forces and the economic sector are experiencing a real revolution.

Relatively recently, the challenge of creating "computer morality," i.e., control systems for robots taking into account ethical issues, including the provisions of the international humanitarian law and the rules of fire, was also undertaken. One of the innovative ideas is research on moral and ethical modules aimed at evoking guilt in robots (Arkin, 2008; Kopeć, 2016). In conditions of peace, especially in laboratories, such projects may turn out to be effective, while in conditions of a network-centric and multilinear war, in which the front lines, lines of demarcation of own troops and, above all, targets remain out of sight, "ethical robots" will probably become useless. Nevertheless, it is a good omen for developing directions of research focused on the culture of waging war without unnecessary casualties. A different issue, which is, at the same time, the most problematic, is the use of AI for risk reduction in the event of a nuclear war (Geist & Andrew, 2018, pp. 1–23). More attention to this issue is given in the next chapter of the article.

On the grounds of the selected concepts and designs of robots presented above, the questions arise: *will and to what extent intelligent machines, in general, dominate the modern battlefield? Is the era of robotization or "military robolution" fiction, or has it begun?* Judging by the pace of development of robotics AI, it is an inevitable process resulting from technological progress. However, there are social and ethical reasons that the elimination of humans from the AI decision loop and the development of the so-called "thinking robots" will be limited. It mainly concerns humanoid robots capable of making independent decisions. The thesis should be treated as a postulate dictated by the good of humanity.

4 Ethics and "Anti-ethics" of Cyberwar

Beyond its purely technological sphere, the robotization of war reveals much controversy. Many doubts are raised by the semantic meaning of the term robot and its attributes, i.e., intelligence and independent thinking. The cognitive sources of these two issues draw from the theses contained in Alan Turing's article from 1950 entitled "Computing Machinery and Intelligence" (1960), whose keynote was to answer the questions "can machine think?", And if so, should they be equipped with communication attributes? The anthropocentric approach to a machine made it necessary to look at it from the perspective of replicating human intelligence to non-human objects, which consequently raised severe ethical controversies. It was based on Turing's considerations that a dilemma appeared in defining the concept of a robot. In the 1950s, the original name was associated with hard work, and the robot itself was a machine (i.e., a technical device performing work based on energy transformations), or, more precisely, a fusion of a machine and a computer (Kopeć, 2016, p. 39). In later years, the critical issue exposed in the definition of a robot was artificial intelligence, which John McCarthy introduced as "the science and engineering of making intelligent machines" into the systematics of concepts as early as 1965 (Peart, 2017). It should be strongly emphasized that AI has become an essential attribute of military robots alongside such features as communication, manipulation, mobility, and thinking based on a three-level architecture. Its first level is based on superficial relationships between visual or auditory stimuli. The second layer allows transforming commands from the deliberative layer into tasks, while the third one is the actual artificial intelligence that enables the robot to plan the next steps (tasks) and collaborate with humans (Lin, Bakey, Abney, 2008, pp. 101–102). This layer gave rise to resolving the dilemma of understanding AI in terms of "thinking" or "intelligence." This dilemma has not been fully resolved to this day. Nevertheless, the most popular interpretation of artificial intelligence at the time was the statement that it was the ability to adapt to new conditions and perform new tasks, see dependencies and relationships, or learn (Różanowski, 2007, p. 110).

The dilemma of the autonomous use of robots in the modern battlefield arises when adopting a modern approach to defining a robot as a machine that feels, thinks, and acts, and, above all, is independent in action. The main problem of the considerations is the asymmetry of opportunities in the confrontation between a human and a machine. It is about the technological dehumanization of war or the phenomenon known as *post-human warfare* (Coker, 2002, p. 399). A human's role in this type of activity is associated with the desire to minimize their own losses and achieve an advantage thanks to the technological factor, i.e., *risk-transfer militarism* (Shaw, 2002). It is limited to setting strategic, operational, or tactical goals and delegating the competence to destroy them to a machine equipped with artificial intelligence.

There are two approaches involved: *targeted killing* and *armchair warfare* (Olson & Rashid, 2013, p. 2). The first one is to eliminate a strategic target without indirect or accidental casualties, which seems to be justified for security and ethics. The latter should be seen as a form of "soft hypocrisy" or "dormant conscience" of people—soldiers giving orders to robots. On the one hand, their autonomous operation, increasing the distance to targets, precision, and digitization of control and evaluation of the task performance effectiveness, contributes to the greater objectification of a human being understood as a goal in hostilities. On the other hand, there is "blurring" consciousness and displacement of the sense of responsibility for the effects of robots' actions. Due to their autonomous functioning and the ability to shape their behavior thanks to software, they are often treated subconsciously as fully thinking and aware of their actions.

On this basis, the "soft hypocrisy" associated with the concept of *human out of the loop*, that is, the elimination of a human from the decision loop in the OODA observation-orientation-decision-action system introduced into the systematics of concepts by John Boyd (Marra & McNeil, 2012), appears. Eliminating a human from the decision loop creates grounds for supposing that, if not now, in the future, the robotization of war will lead to the disappearance of any ethical values on the part of soldiers—commanders, and a human will become a target devoid of the human being's attributes. It relates to the role of a human being at the stage of setting a target or determining a task for robots or precision-guided munition (PGM) missiles. These weapons use a system that seeks and track a target thanks to energy reflecting the target or aiming point and then processing data from reconnaissance into commands that direct weapons to the target (Puckett, 2004, p. 648).

The use of robots and precision (homing) ammunition is associated with the accuracy of reconnaissance and identification of a target as an enemy, not a defenseless civilian or soldier of own troops. The autonomous operation of robots or the homing of intelligent missiles can lead to serious mistakes, the commanders' responsibility for which is repeatedly blurred. The very name of this type of fire and forget weapon raises ethical controversy. It indicates the blurring of the so-called "emotional bond" between an attacker and an opponent. D. Grossman emphasizes this phenomenon more clearly, pointing to the multidimensional "separation" of a soldier from killing (in the spatial, temporal, and even moral dimension—by shifting decisions onto autonomous machines). According to the author, this will avoid the aversion to killing, which is an essential factor that reduces effectiveness, especially when the soldier observes their victim directly (Grossman, 2010).

However, there are ethical reasons for the implementation of intelligent robots in wars. The first one is the reduction of own losses, especially among soldiers, supported by the data from the 1991 Gulf War when the use of precision weapons enabled coalition troops to suffer minor losses to the forces and resources used. The US. Army lost 355 killed, including 148 in combat, 496 wounded in combat, and around 3000 soldiers off the battlefield. There were 47 killed and 43 wounded on the British side, the French—2 killed and 25 wounded, the Egyptian - 14 and 120, respectively. From 20 to 35 thousand died, and from 25 to 60 thousand soldiers were wounded on the Iraqi side (Biziewski, 1994). The war was assessed as a turning point in the development of precision weapons and the beginning of artificial intelligence advancement on the battlefield. The lessons were learned from it and fully used in the conflict in the Balkans 1991–1995, Afghanistan in 2001–2020, Iraq in 2003, and the war in Syria that continues to this day. The second premise for the robotization of the war is the partial elimination of soldiers' fear of losing their lives, as well as the effects of war in the post-traumatic stress disorder (PTSD) (Netczuk-Gwoździewicz, 2020). Lack of contact with the enemy and the effects of using conventional means creates a sphere of a specific type of comfort among soldiers taking part in hostilities. In most cases, the comfort is apparent and can be treated as "a credit of conscience with a deferred grace period."

Not only ethical controversies are raised by the concept of using AI as a system for supporting or controlling nuclear weapons. It is currently solely dedicated to the security systems of nuclear missile installations and detecting and neutralizing foreign nuclear missiles. In the future, it is expected to be employed more widely, namely thanks to "superintelligence," it may become an autonomous system to respond to nuclear attacks carried out by the enemy. The concept of "superintelligence" has as many enthusiasts as opponents since it carries the risk of an uncontrolled war on a global scale. Experts indicate that AI should, of course, be used, but as an element of decision-making support, correcting human errors or identifying threats of the enemy's use of nuclear weapons (Geist & Andrew, 2018, p. 2). The fundamental problem of AI usage in combination with nuclear weapons is the susceptibility of the latter to cyber-attacks related to the insufficient level of security of information systems and the lack of knowledge about vulnerabilities in such systems in other armed forces (Avin & Amadae, 2019). This means that the implementation of autonomous, intelligent nuclear weapons command systems may lead to the extermination of humanity (Boulanin, 2019, pp. 113–115). Currently, the US recommends avoiding autonomous nuclear security and command systems for two reasons. The first relates to the cyber vulnerability of own missile installations to hacker attacks. The other one concerns similar problems with potential opponents. Nonetheless, the overriding political and ethical problem seems to be the threat of terrorists or criminal groups taking control of the nuclear potential, or worse by AI itself.

AI combined with nuclear weapons is not only the US domain. In November 2015, Russia revealed that it was developing the final version of the "Killer Robot," a nuclear-powered underwater drone "Status-6," designed to carry a powerful thermonuclear warhead. This weapon was still shrouded in mystery in the year under review, but the public learned of its existence thanks to Russian television (according to unofficial sources deliberately leaked). The advantages of the weapon are elements such as AI-guided firepower and speed that allows performing the task independently. Considering its capabilities, which include a stealthy approach to an enemy target, speed of movement, and the ability to strike with a force that guarantees the destruction of thousands of people, the weapon should be classified as extraordinarily barbaric and inhuman. Although it is only a concept, it opens a new chapter in research into the use of AI in a future war (Boulanin, 2019).

The presented examples draw attention to philosophical approaches to AI in the present and future wars. There are many conflicting positions on this point. One opponent of using AI-equipped combat robots is R. Sparrow, who argues that any use of "fully autonomous" robots is unethical due to the *Jus in Bello* requirement or liability for possible war crimes. His position is based on deontological and consequentialist arguments embedded in normative ethics (Sparrow, 2006; Arkin, 2011, p. 8). In this sense, deontology is concerned with a set of ethical theories that there are certain kinds of acts that must not be done, even if it leads to worse results. In other words, according to deontologists, the moral value of acts, principles, laws, or character traits is determined not only by their effects but also by their internal value, which may be influenced, for example, by intentions, motivations, or the type of the assessed act or principle. On the other hand, consequentialism is the name of a set of ethical theories that state that the non-instrumental, internal moral value of acts, rules, laws, character traits, or institutions have only their effects.

The use of AI in war violates the basic principles of the commander's liability for deciding for its use and the consequences of its autonomous operation (Sparrow, 2006). At the other end of the debate, there is the thesis that the level of technological war advancement and the pace and unpredictability of warfare even impose the use of autonomous systems equipped with AI. Other ethical considerations hint to an analogy between robot warriors and child soldiers, as neither is responsible for their own decisions (Arkin, 2011, p. 8). Accountability appears to be one of the main ethical and legal concerns, but it is not the only problem.

The ethical counterargument to Sparrow's views is the question of the nature of a human, their tendency to unethical behavior, fallibility, and spontaneity raised earlier. Thus, assuming a human being able to create an "ethical robot" with a full package of programmed data and pro-social skills, it may turn out to be the equivalent of a soldier on the battlefield capable of performing tasks under the provisions of the international humanitarian law concerning armed conflicts and ethics (Arkin, 2011, pp. 8–9). It is about a robot with "moral intelligence," currently being researched in the USA and Japan. That is also evidenced by the ability to maintain conservative robots, devoid of self-preservation and fear for one's own life, as well as the lack of emotions in action (Walzer 1977, s. 251). Innovative and bold theses show that a robot can be responsible for its own actions. One of the precursors of this approach is J. Sullins, who argues that robots can be ascribed moral agency if they have intelligence, programmed morality, and the attributes of a thinking being (2006). This approach should be treated as quite brave and controversial since legal sanction and social control are related to a simple question—*how to punish a robot*? Equally controversial in the philosophical debate is the notion of risk-free war, equated with the notion of just war. (Walzer, 2004, p. 16).

These and many other philosophical approaches to the use of intelligent robots in war, even if they are abstract and unrealistic for practical implementation, indicate that the ethical problems of using AI on the modern battlefield have not escaped the attention of representatives of the world of science, politicians, or military commanders. Its development is the result of the conceptual struggle and analysis in the case study of conflicts and wars conducted by the US using AI. The very idea of the document is the result of deeply entrenched values such as leadership, professionalism and technical knowledge, dedication to duty, honesty, ethics, honor, courage, and loyalty, and compliance with the provisions of the international humanitarian law. The essence of the document is the ethical principles proposed as the canon of the culture of warfare conducted by the USA. The first principle is the **responsibility**, which, apart from the classic provisions relating to commanders and executors' legal liability, categorically rules out the possibility of delegating responsibility to "machines" equipped with AI. The principle of responsibility is based on the theory of consent. Such consent requires that individuals meet three criteria: the action is voluntary, deliberate, and informed (DiB, 2019; Simmons, 1981). The second rule—**equity**—establishes that systems equipped with AI cannot cause any unnecessary damage. The third principle speaks of **traceability** of AI systems, understood as the possibility of controlling and auditing its functioning at every stage of operations. It aims to limit the AI taking control of the system/weapons. Another principle is **reliability**, based on a clearly defined field of application (AI purpose), security, protection, and robustness of systems equipped with AI. A critical element of the system reliability is its resistance to failures and the possibility of their removal at every operation stage. The fifth principle—**governability**—assumes that systems equipped with AI should have the ability to interrupt operation, isolation, and self-control safely. They should be designed and constructed to fulfill their intended function while having the ability to detect and avoid unintentional damage or disturbance and human or automated withdrawal or deactivation of use (DiB, 2019). The added value of the document is a set of recommendations for users of AI-equipped systems.

5 Conclusions

The presented issues merely delineate some areas specific to the debate on the ethics of AI use on the modern battlefield. These issues form a platform for further considerations, especially about the "modern culture of killing" presented in the title. Its essence is revealed not so much in the increasing physical distance between the decision-maker – a commander and the task performer—a "machine" equipped with AI, but in the mental distance revealed in the disappearance of the sense of responsibility for physical and moral war damage. The dehumanization of war connects both concepts, the so-called post-human warfare, which is symbolically described by two approaches: targeted killing and armchair warfare. While the first approach is legitimate as it aims to eliminate strategic goals without accidental casualties, the second should be viewed as a form of "soft hypocrisy" or "dormant conscience" of the commanders. The metaphorical approach to the war waged from behind a computer screen based on a virtual mapping of targets (including people) reduces the act of death to the level of a computer game in which everything is permissible, and the only form of responsibility is the quantification (statistics) of the victims of the decision made.

Considering the research question on the social (ethical) effects of using AI in contemporary conflicts and wars, it should be presumed that its application is still a matter that requires not only public debate but also scientific research on the elimination or limitation of the human role from the AI decision loop. Particularly significant problems that need to be resolved are research on autonomous robots equipped with "superintelligence" that exclude humans from the decision loop and the concepts of "computer morality," i.e., the artificial application of ethics in war and the provisions of the international humanitarian law. In the opinion of ethics, theologians, psychologists, and sociologists, both research lines, although very abstract and even unreal, lead to an ethical utopia.

The considerations taken allow the statement that AI development combined with modern military technology has already influenced the objectification of a human understood as a goal—material or virtual object to be liquidated without a human being's attributes. The existence of a "modern culture of killing" is justified by many premises, the most vital of which is related to the loss of emotional ties between the decision-maker issuing orders/commands and AI-equipped robots or systems, and the loss of the sense of responsibility for the order issued. Thus, the "modern culture of killing" means the philosophy of waging a humanitarian war through the use of modern military technologies.

References

Antal, J. (2016). The next wave racing toward robotic war. *Military Review*. https://en.topwar.ru/101998-sleduyuschaya-volna-naperegonki-k-voynam-robotov.html (2020.09.30).

Arkin, R. (2011). *Governing lethal behavior: Embedding ethics in a hybrid deliberative/reactive architecture*. Georgia Institute of Technology, http://www.cc.gatech.edu/ai/robot-lab/online-publications/formalizationv35.pdf. (2015.10.30).

Avin, S., & Amadae, S. M. (2019). *Autonomy and Machine Learning as Risk Factors at the Interface of Nuclear Weapons, Computers and People, The Impact of Artificial Intelligence on Strategic Stability and Nuclear Risk: Euro-Atlantic Perspectives*, SIPRI, May. 105–118.

Biziewski, J. (1994). *Pustynna Burza*. Część 2, [Desert Storm. Part 2.], Warszawa: Altair.

Bendett, S. (2018). Here's how the Russian military is organizing to develop AI, *Defense One*, July 20, 2018. https://www.defenseone.com/ideas/2018/07/russian-militarys-ai-development-roadmap/149900/.

Bendett, S. (2017). Red robots rising: Behind the rapid development of Russian unmanned military systems, *The Strategy Bridge*, December 12, 2017, https://thestrategybridge.org/the-bridge/2017/12/12/red-robots-rising-behind-the-rapid-development-of-russian-unmanned-military-systems.

Bird, E., Fox-Skelly, J., Jenner, N., Larbey, R., Weitkamp, E., & Winfield, A. (2020), *The Ethics of Artificial Intelligence. Study Panel for the Future of Science and Technology*, EPRS-European Parliamentary Research Service, Scientific Foresight Unit (STOA), PE 634.452 – March 2020.

Bostrom, N., & Yudkowsky, E. (2014). *The ethics of artificial intelligence*. In K. Frankish, W. Ramsey (Eds.), *Cambridge handbook of artificial intelligence* (pp. 314–334). New York: Cambridge University Press.

Boulanin, V. (2019). *The impact of artificial intelligence on strategic stability and nuclear risk*, Vol. I, Euro-Atlantic Perspectives, SIPRI.

Boussios, E. G., & Visvizi, A. (2017). Drones in War: The controversies surrounding the United States expanded use of drones and the European Union's disengagement. *Yearbook of the Institute of East-Central Europe, 15*(2), 123–145.

Coeckelbergh, M. (2011). Artificial companions: Empathy and vulnerability mirroring in human-robot relations. *Studies in Ethics, Law and Technology, 4*(3), 1–17.

Coker, Ch. (2002). Towards post-human warfare: Ethical implications of the Revolution in Military Affairs, Die Friedens-Warte. *Journal of International Peace and Organization*, nr 77/4.

Cummings, M. L. (2017). *Artificial Intelligence and the Future of Warfare, International Security Department and US and the Americas Programme January 2017*, Chataman House, The Royal Institute of International Affairs.

Danet, D., Hanon, J.-P. (2014). *Digitization and robotization of the battlefield: evolution or robolution?* In R. Doaré et al. (Eds.), *Preface, in: robots on the battlefield, contemporary issues and implications for the future*. Fort Leavenworth.

Darling, K. (2016). *Extending legal protection to social robots: The effects of anthropomorphism, empathy, and violent behavior towards robotic objects*, Calo, Froomkin.

De Spiegeleire, S., Maas, M., & Sweijs, T. (2017). *Artificial intelligence and the future of defense strategic implications for small- and medium-sized force providers*. The Hague Centre for Strategic Studies, Hague.

DIB (2019)*AI Principles: Recommendations on the Ethical Use of Artificial Intelligence by the Department of Defense*, Supporting document, Defense Innovation Board (DIB), October 2019.

European Commission (2019) *A definition of AI: Main Capabilities and Scientifi c Disciplines.* (2019). High-Level Expert Group on Artifi cial Intelligence, Brussels.

Frey, C. B., Osborne, M. A. (2013). *The future of employment: How susceptible are jobs to computerisation?* Oxford Martin Programme on the Impacts of Future Technology.

Geist, E., & Andrew, J. L. (2018). *How might artificial intelligence affect the risk of nuclear war?* RAND Corporation.

Goertzel, B., & Pennachin, C. (Eds.). (2007). *Artificial general intelligence. Cognitive technologies.* Springer.

Green, B. P. (2016). Ethical reflections on artificial intelligence. *Scientia Et Fides, 6*(2), 9–31.

Grossman, D. (2010). *O zabijaniu. Psychologiczny koszt kształtowania gotowości do zabijania w czasach wojny i pokoju (About killing. The psychological cost of shaping the willingness to kill in times of war and peace)* (pp. 185–189), Warszawa.

Grosz, B. J., Russ, Ch., Altman, Horvitz, E., Mackworth, A., Mitchell, T., Mulligan, D., & Shoham, Y. (2016).*One hundred year study on artificial intelligence, artificial intelligence and life in 2030.* Report of the 2015, Study Panel Stanford University, September 2016.

Hinton, G. E., & Salakhutdinov, R. R. (2006). Reducing the dimensionality of data with neural networks. *Science, 313*(5786), 504–507.

IFR (2008) *World Robotics Report 2008*, Statistical Department, Frankfurt am Main, Germany.

IFR (2020) *World Robotics Report 2020*, Statistical Department, Frankfurt am Main, Germany.

Kamieński, Ł. (2014). *Nowy wspaniały żołnierz. Rewolucja biotechnologiczna i wojna XXI wieku* (Great New Soldier. The Biotechnological Revolution And The War Of The 21st Century). Kraków, Wydawnictwo Uniwersytetu Jagiellońskiego.

Kopeć, R. (2016). Robotyzacja wojny (Robotization of war). *Społeczeństwo i Polityka, Nr, 4*(49), 37–53.

Leslie, D. (2019). *Understanding artificial intelligence ethics and safety A guide for the responsible design and implementation of AI systems in the public sector*. Public Policy Programme.

Lin, P., Bakey, G., & Abney, K. (2008). *Autonomous military robotics: Risk, ethics, and design.* Waszyngton.

Lucas, G. R. (2014). Automated warfare. *Stanford Law & Policy Review, 25*, 317–339.

Luttwak, E. N. (1995). Toward post-heroic warfare: The obsolescence of total war. *Foreign Affairs, Nr, 74*(3), 109–122.

Luxton, D. D. (2014). Artificial intelligence in psychological practice: Current and future applications and implications. *Professional Psychology: Research and Practice, 45*(5), 332–339.

Madiega, T. (2019). *EU guidelines on ethics in artificial intelligence: Context and implementation.* EPRS, European Parliamentary Research Service, Members' Research Service, PE 640.163 – September.

Malik, R., Visvizi, A., & Skrzek-Lubasińska, M. (2021). (2021) The gig economy: Current issues, the debate, and the new avenues of research. *Sustainability, 13*, 5023. https://doi.org/10.3390/su1 3095023

Marra, W. C., & McNeil, S. K. (2012). Understanding "the Loop": Regulating the next generation of war machines. *Harvard Journal of Law & Public Policy, Nr, 36*, 1139–1185.

Mike, M. W., & Schinzinger, R. (2010). *Introduction to engineering ethics* (2nd ed.). McGraw-Hill.

Minsky, M. (1961). Steps toward artificial intelligence. In *Proceedings of the IRE* (Vol. 49, No. 1, pp. 8–30).

Mearsheimer, J. (2001). *The tragedy of great power politics*. W. W. Norton & Company.

Morgenthau H. J. (1973). *Politics among nations. The struggle for power and peace*. New York: Alfred A. Knopf.

Netczuk-Gwoździewicz, M. (2020), *Psychologiczne uwarunkowania bezpieczeństwa żołnierzy – uczestników misji poza granicami kraju [Psychological conditions for the safety of soldiers - participants of missions abroad]*. Wrocław: Wydawnictwo AWL.

Nye, J. S. (2011). *The Future of Power*. Public Afairs.

Olson, J., & Rashid, M. (2013). *Modern Drone Warfare: An Ethical Analysis, American Society for Engineering Education Southwest Section Conference*, http://se.asee.org/proceedings/ASE E2013/Papers2013/157.PDF (30.10.15).

Özdemir, S. G. (2019). *Artificial intelligence application in the military the case of United States and China*, SETA Analysis No.51, Istambul, Ekonomi Ve Toplum Araştirmalari Vakfi.

Pathe Duarte, F. (2020). Non-kinetic hybrid threats in Europe—The Portuguese case study (2017–18). *Transforming Government: People, Process and Policy, 14*(3), 433–451. https://doi.org/10. 1108/TG-01-2020-0011

Peart, A. (2017). Homage to John McCarthy, the father of Artificial Intelligence (AI), https://www.artificial-solutions.com/blog/homage-to-john-mccarthy-the-father-of-artificial-intelligence (2019.09.06).

Puckett, Ch. B. (2004). In this era of "smart weapon". Is a state under an international legal obligation to use precision-guided technology in armed confl ict. *Emory International Law Review*, nr 18/2.

RAND. (2019). *The Department of Defense Posture for Artificial Intelligence* (p. 2019). RAND Corporation, Santa Monica.

Remmers, P. (2019). The ethical significance of human likeness in robotics and AI. *Ethics in Progress, 10*(2), 52–67. Art. #6.

Różanowski, K. (2007). Sztuczna inteligencja: rozwój, szanse i zagrożenia (Artificial intelligence: development, opportunities and threats), *Zeszyty Naukowe Warszawskiej Wyższej Szkoły Informatyki*, nr 2.

Sayler, K. M. (2020). *Artificial intelligence and national security*. Congressional Research Service. R45178, version 8.

Safder, I., UlHassan, S., Visvizi, A., Noraset, T., Nawaz, R., & Tuarob, S. (2020). Deep learning-based extraction of algorithmic metadata in full-text scholarly documents. *Information Processing and Management, 57*(6), 102269. https://doi.org/10.1016/j.ipm.2020.102269

Sharkey, N. (2008). Computer science: The ethical frontiers of robotics. *Science, 322*(5909), 1800–1801.

Shaw, M. (2002). Risk-transfer militarism, small massacres and the historic legitimacy of war. *International Relations, nr 16*(3), 343–360.

Simmons, A. J. (1981). *Moral principles and political obligations*. Princeton University Press.

Sparrow, R. (2006). Killer robots. *Journal of Applied Philosophy, 24*(1).

Sullins, J. (2006). When is a robot a moral agent? *International Journal of Information Ethics, 6*, 12.

Troisi, O., Visvizi, A., & Grimaldi, M. (2021), The different shades of innovation emergence in smart service systems: the case of Italian cluster for aerospace technology. *Journal of Business & Industrial Marketing*, Vol. ahead-of-print https://doi.org/10.1108/JBIM-02-2020-0091

Tonin, M. (2019). *Artificial intelligence: Implications for Nato's armed forces*, Science And Technology Committee (STC), Sub-Committee on Technology Trends and Security (STCTTS), 149 STCTTS 19 E rev. 1 fin.

Turing, A. (1960). Computing machinery and intelligence. Mind, nr 59/236, 433–460.

Visvizi, A., & Lytras, M. D. (Eds.). (2019). *Politics and technology in the post-truth era*. Emerald Publishing.

Visvizi, A., Lytras, M. D., & Daniela, L. (2020). *The future of innovation and technology in education: Policies and practices for teaching and learning excellence*. Emerald Publishing,

Walzer, M. (1977). *Just and unjust wars*, 4th edn. Basic Books.

Walzer, M. (2004). *Arguing about war*. Yale University Press.

Windeck, A. (2014). Preface. In R. Doaré, D. Danet, Jean-Paul Hanon, & G. de Boisboissel (Eds.),*Robots on the battlefield, contemporary issues and implications for the future*. Combat Studies Institute Press, US Army Combined Arms Center, Fort Leavenworth, Kansas.

Wong, Y. H., Yurchak, J. M., Button, R. W., Frank, A., Laird, B., Osoba, O. A., Steeb, R., Harris, B. N., & Bae, S. J. (Eds) (2020). *Deterrence in the age of thinking machines*. RAND Corporation, Santa Monica, California.

Marek Bodziany Ph.D. in Sociology, Associate Professor, Colonel of the Polish Armed Forces, Deputy Dean of the Faculty of Security Sciences at the Military University of Land Forces, Chief Editor of the AWL Publishing House. Research areas: multiculturalism, education, culture sociology, ethnical and cultural conflicts, wars and social crises, methodology, migration and social change. Email: m.bodziany@interia.pl ORCID: 0000-0001-8030-3383

Security Dimensions of AI Application

Stratcom & Communication Coherence in the Polish Armed Forces

Piotr Szczepański

Abstract In context of the military, StratCom cells are responsible for integrating communication effort and coordinating activities in the field of communication capabilities, functions, and tools, including psychological operations (PsyOps), information operations (Info Ops), and military public affairs (MPA). Every day, when observing, for example, image crises, one can see how vital and critical links are caused by incompetent information management. The objective of this chapter is to examine the centrality of StratCom for the success of military operations. Against the backdrop of the Polish Ministry of National Defense StratCom, a detailed insight into diverse areas of StratCom is provided. After considering the modern requirements of the information environment, factors influencing the StratCom process are be identified. Discussion and recommendations follow.

Keywords StratCom (StratCom) · Information operations (InfoOps) · Artificial Intelligence (AI)

1 Introduction

StratCom is an area of actions taken by the armed forces, which greatly facilitate the performance of the tasks set. It concentrates processes and efforts undertaken to understand and engage key audiences to create, strengthen, or consolidate conditions favorable to the implementation of national interests and goals using coordinated information, topics, plans, programs and activities synchronized with projects implemented by other elements of state authorities (DoS, 2006, p. 3). This concept, derived from the theory and practice of civilian organizations' functioning, found its reflection in the armed forces of many countries. The United States (US) was a pioneer in this field, but other member states of the North Atlantic Treaty (NATO) soon turned their attention to this issue too.

The definitions of StratCom pointed to the dualism of the concept in terms of both the information message and actions taken to shape the image of the state (Josten,

P. Szczepański (✉)
Military University of Land Forces, Wrocław, Poland

© The Author(s), under exclusive license to Springer Nature Switzerland AG 2021   65
A. Visvizi and M. Bodziany (eds.), *Artificial Intelligence and Its Contexts*, Advanced Sciences and Technologies for Security Applications,
https://doi.org/10.1007/978-3-030-88972-2_5

2006, p. 16) as well as the potential possibilities related to the implementation of political assumptions, the decision-making process, the dissemination of credible information on the decisions taken and the actions that result from them, as well as extremely important, culturally adapted communication with the audience (Stavridis, 2007, p. 4).

The definition of information operations (actions) adopted for the StratCom purposes was reflected in allied publications, the most important of which is NATO Policy on Information Operations. It is approached as a military function that provides advice and coordination of military actions aimed at achieving the desired effect in the sphere of will to act (fight), perception, and ability to act by the enemy, potential adversary, and other parties to the conflict approved by the North Atlantic Council in support of the objectives of operations conducted by the Alliance (NATO, 2007, p. 2). The US Department of Defense has included the following elements in StratCom: Information Operations, Public Diplomacy, International Broadcasting Services, and Public Affairs (DoD, 2004).

The main discrepancy in the presented topics is the current state of knowledge about the methods, tools, and activities used by the Polish Ministry of National Defense, affecting the ability to implement StratCom and their effectiveness in achieving the set goals. Recognition of the methods and tools used so far, an indication of those that are to direct the Polish Armed Forces' activities appropriately, and finally, the development of recommendations in effective StratCom determined the setting of the research problem. The objective of this chapter is to query to what extent the Ministry of National Defense can implement StratCom at all levels of command. The objective of this chapter is to examine the question of coherence of communication processes at all command levels in the armed forces in the field of StratCom (*StratCom*), including information operations (*Info Ops*). The argument in this chapter is structured as follows. The following section, by means of an introduction to the topic, offers a literature review. Here the concepts of StratCom, including InfoOps, and other issues are brought to the surface of the discussion. Then, a brief overview of threats in the information environment follows. In the next step, the vital role of Artificial Intelligence (AI) in modern StratCom is discussed. Discussion and conclusions follow.

2   Literature Review

The dynamically changing information environment poses several new challenges for the armed forces. The current perception of the environment of the activities carried out and the communication undertaken with recipients who are an integral part of this environment became insufficient in the dynamic development of information and communication technologies and cultural and civilization changes. The NATO has experienced it while implementing tasks in various regions of the world. The experience resulted in the initiation of work on developing the Alliance's StratCom concept, which brought the publication of numerous related documents.

The *NATO Policy on StratCom* (NATO, 2009), was of crucial importance in developing the StratCom concept. By issuing it, the Alliance reaffirmed that it recognizes the importance of an appropriate, timely, accurate, and active way of communicating its evolving role, objectives, and subordinated tasks. StratCom has been recognized as an integral part of the Alliance's efforts to achieve its political and military goals.

Initiating the process of modifying the StratCom concept, especially in the aspect of tasks resulting from the scope of individual structural bodies' responsibilities, led to the publication of another important document—*Allied Command Operations Directive on StratCom* (SHAPE, 2009). It contained guidelines for planning and conducting StratCom projects within the structures subordinate to Allied Command Operations (ACO). It does not change the basic assumptions in the *NATO Policy on StratCom* but it complements, specifies, and clarifies them.

Finally, the *2010 NATO Military Concept for StratCom* developed by the Allied Command Transformation (ACT, 2010) should be mentioned. The concept summarizes the StratCom assumptions developed so far. It defines the link between StratCom and the North Atlantic Council's information strategy. It points to the coordinating nature of Info Ops and, at the same time, emphasizes the afore-mentioned separateness of these operations and press and information activities. The importance of leadership and all-level commanders' responsibility for proper communication with the objects of influence and the need to modify the existing organizational structures are also noticed.

Activities in the broadly understood information sphere were carried out practically from the beginning of military operations history (JFSC, 2004). A breakthrough in developing new information activities (operations) was the first war in the Persian Gulf (1990–1991). At that time, the unique role of information activities, including psychological activities, was noticed. The concept of information operations that emerged then was reflected in the subject literature. They are devoted to the information aspect of the completed campaign and analyzes the impact of information and the role of information systems in conducting current military operations (Campen, 1992; Resteigne & Bogaert, 2017; Soeters, 2018; Johnsen & Chan, 2019; Boussios & Visvizi, 2017). Military specialists were working on this subject at the same time, which resulted in the publication of *Memorandum of Policy (MOP) No 30" (Command and Control Warfare)* by the US College of Chiefs of Staff in 1993. The Memorandum defined the method for influencing the command systems (Command and Control Warfare—C2W) systems of a potential enemy while defending them by their own troops.

As for the development of the concept of information warfare, it is not a separate way of waging war. It is a mosaic of various forms of warfare in the information sphere (Libicki, 1995; Szpyra, 2003, p. 13), including: a struggle for an advantage in command, combat in the sphere of reconnaissance, electronic combat, psychological struggle, hacking struggle, economic struggle, and cyber struggle. It is particularly essential for the definition of information operations and the elements included in them. The information warfare covers the entire spectrum of military operations and all levels of combat (Fredericks, 1997, p. 98).

A breakthrough event resulting from the publication mentioned above was a discussion leading to the conclusion that the term information warfare is not adequate to the scope of designations assigned to this concept. That is, since a wide range of undertakings in the information sphere is also carried out in peacetime, it is difficult to classify them as a fight (Szpyra, 2003, p. 92). Hence, it was reasonable to find a more appropriate term for this kind of action. Accordingly, it was agreed that the definition of *Information Operations* would be the most appropriate (DA, 1996; JCS, 1998; USAF, 1998).

Currently, the U.S. Department of Defense defines these activities as the integrated use of critical electronic warfare capabilities, computer network operations, psychological activities, disinformation, and the safety of activities synchronized with specific support and related capabilities that are implemented to impact, disrupt, damage, or take control of the opponent's decision-making process, both human-made and automated, while protecting one's potential in this regard (JCS, 2006, p. IX). Partly in parallel with the development of the allied StratCom concept, and sometimes due to the emergence of this concept, individual Member States began to develop their own national solutions in this area.

2.1 Documents Related to the StratCom in the Polish Armed Forces

The Polish StratCom concept does not directly reflect the allied concept, but is a proposal to implement a coherent system solution. As for the StratCom and Info Ops areas the following documents are currently in force at individual Polish Armed Forces command levels:

– *Decision No. 478/MON of December 8, 2014 on the StratCom System in the Ministry of National Defense*—document implemented (MON I, 2014);
– *Information Operations*—(MON VII, 2017)—the document was introduced in the Polish Armed Forces on October 2, 2017 with the deadline for its implementation by September 31, 2018;
– *Psychological Operations*—(MON II, 2017)—document implemented;
– *Radioelectronic Warfare*—(MON VI, 2015)—document implemented;
– *Operational Masking*—(MON IV, 2018)—the document has been in force in the Polish Armed Forces from April 1, 2018;
– *Planning Information Activities at the Tactical Level in the Land Forces*—(MON III, 2017)—the document introduced from June 1, 2017 to be used in the General Command of the Polish Armed Forces and subordinate units;
– *Instructions for the Organization and Operation of StratCom within the Military Command System*—2014—the document in force in the Polish Armed Forces (SG, 2014).

Decision No. 478/MON is the only document regulating StratCom in the Ministry of National Defense. The main document in the Info Ops area in the Polish Armed Forces is the doctrinal document *Information Operations* (MON VII, 2017), which has been in force since October 2, 2017. It is necessary to ensure consistency with NATO works, update the existing documents, or develop new supporting ones in this area. The provisions of the document *Planning Information Activities at the Tactical Level in the Land Forces*—(MON III, 2017) should be adapted to the applicable doctrinal document *Information Operations* (MON VII, 2017). Moreover, drawing up supplementary documents concerning the planning of Info Ops activities in individual branches of the Polish Armed Forces should be considered.

Due to the lack of a document on the security of operations and disinformation, based on the draft document developed at NATO—*Allied Joint Doctrine for OPSEC and Deception* (NATO, 2018), provisions from this area were included in the doctrinal document *Operational Masking* (MON IV, 2018) which has been in force in the Ministry of National Defense since April 1, 2018.

The doctrinal document, DD-3.10 (A), defines the essential tasks and competencies at various leadership and command levels. According to DD-3.10 (A), the communication process consists of the analysis of the information environment ensuring the development and systematic updating of databases, planning communication activities at all command levels, and implementing communication activities and their evaluation.

The lack of the Info Ops structure in the Polish Armed Forces results in the non-existence of specific provisions in documents related to Info Ops tasks at individual control and command levels, especially at the tactical level.

The above documents mark the beginning of the process of introducing system solutions. The subsequent stages of implementing StratCom should include developing relevant executive documents, establishing task (functional) structures on their basis, and then their vertical and horizontal integration within the control and command structures existing in the Ministry of National Defense. A vital element of the entire process will be assessing the effectiveness of the adopted solutions and, if necessary, their further modification based on the current conditions of the functioning of the Armed Forces.

3   Examination of the Problem

The new NATO military policy on StratCom MC0628 does not impose any solutions on allied countries. However, it does suggest specific tasks for them. Therefore, one should strive for the convergence of the effect and not structural solutions. When recommending changes, one must consider the national specificity and the specificity of the Ministry of National Defense. Solutions should be introduced to ensure effective integration of the lines of communication effort at all Ministry levels and the coordination of elements of the Polish Army to achieve uniformity of communication and synchronization of communication capabilities and tools.

The Armed Forces of the Republic of Poland can use the experience of individual NATO member states that have developed StratCom structures and confronted their solutions as part of expeditionary and national missions or daily communication activities.

Mutual understanding of needs and achievement of strategic goals at the level of NATO—a member state, considering national interests, should be the basis for cooperation within the Alliance. Cooperation allows learning from Alliance members' experiences in which the areas in question effectively carry out their tasks. Joint seminars, conferences, exercises, or working groups are also an opportunity to share experiences. A good platform for exchanging views may be the NATO StratCom CoE in Riga, of which Poland is a founding country and one of the framework countries.

When studying StratCom issues, it is necessary to learn about the threats present in the information environment.

According to the *Doctrine of Information Operations*—DD 3.10 (A), threats in the information environment are defined as *destructive impacts on entities (political leaders, military commanders, command and communication systems, etc.) carried out by a potential enemy* (DMON VII, 2017, pp. 1–5).

Image threats constitute an essential group in the activities of the armed forces of any country. The most dangerous one is when the planned operational activities will have a negative communication effect (e.g., civilian casualties in the effect of an operation). This type of threat also includes those that arise from enemy actions.

Disinformation and propaganda can hinder the decision-making process and impact military commanders at various levels, all this to undermine leaders' credibility, even incapacitate the armed forces command system, or influence the society or social groups in the media. The consequence is also a reduction in the armed forces' credibility due to planned communication activities.

Their elimination or minimization is achieved by implementing StratCom guidelines and appropriate coordination of activities, communication capabilities, and tools as well as military actions having an indirect impact on the information sphere. Shaping the awareness of commanders at all levels in terms of StratCom skills is also not without significance.

Another group are cyber threats that should be understood as the enemy's influence on infrastructure elements constituting communication channels with recipients (e.g., servers, networks, radio, television) or directly on the critical military infrastructure to disrupt the decision-making process. The task of this type of threat is temporary or permanent damage to networks, ICT systems, or devices through which communication is carried out both inside and outside the Ministry. A frequent practice is the acquisition of social media profiles, which are used to communicate with audiences as part of broader StratCom. That can negatively affect the institution's image and become a manipulation tool. In this case, there is a risk of stealing sensitive data remaining within the ICT systems to impersonate persons from the attacked institution.

Information noise is also likely to occur and result in a lack of coherent message, which is crucial for StratCom.

Adequate protection of the infrastructure necessitates maintaining appropriate security standards at all command levels. Therefore, an essential element in the StratCom system is the protection of the national cyberinfrastructure.

The category of threats to the information environment occurring during an operation or war are threats of war. Their main task is to disrupt the commanders' decision-making process and lower the soldiers' morale and their will to fight. In this case, the primary modus operandi is to sabotage and destroy communications, intelligence, and information systems.

War threats in the information environment include events occurring when using the full range of communication activities is impossible. It stems from legal and ethical limitations regarding the use of forces and means of the destruction system in peacetime. Such a situation may be implementing offensive communication activities by the enemy already in peacetime to weaken the state's capabilities in, for example, public diplomacy, management, and command of the Polish Armed Forces, including the implementation of military StratCom tasks. The opposing party will impersonate government officials, discredit forces, and resources within the national defense system, manipulate information, and conduct propaganda and disinformation activities.

Artificial intelligence plays a vital role in the description of modern StratCom. AI technology that already exists promises significant benefits. AI promises to help public administration, enabling the state to deliver more and better services cheaply through improved anomaly detection, demand prediction, tailoring services, and more consistent and scalable decision-making in general. Building on behavioral insights and commercial targeting techniques, AI can even be employed to influence citizens' behavior in ways that promote social objectives (Ranchordás, 2019).

Recent years have seen significant advances in the use of AI to diagnose certain conditions-especially those relying on the analysis of medical scans-opportunities that are vital to seize. Nevertheless, online platforms are incentivized to hoover up increasing amounts of our data, often in ways that are not transparent. The ubiquity of and access to social media, the windfall of big data, and the resulting advances in analytics and artificial intelligence have disrupted traditional approaches.

Artificial intelligence tools, such as bots deployed on social media platforms, can amplify narratives for online audiences beyond what they would ordinarily achieve in real-world settings. That makes it possible for individuals holding similar views to converge on an idea, reinforce their beliefs, and drown out contrary views. It enables disinformation campaigns by vested interests to travel at speed, leaving little room for rebuttal and retraction (Allen, 2020).

The rapid development of technology has dramatically changed the information environment. The NATO Military Policy of Information Operations (NATO, 2015) defines the Information Environment as one that is comprised of the information, individuals, organizations, and systems that receive, process, and convey information, and the cognitive, virtual, and physical space in which this occurs. In this context, a crucial aspect is the media monitoring process, which allows for a quick response after detecting signals that may negatively impact the Armed Forces' image.

In the Polish Armed Forces, media monitoring is carried out by the Operational Center of the Ministry of National Defense and the Monitoring and Analyzes Center and the Central Group of Psychological Activities. It is primarily about collecting information from the media regarding the Polish Armed Forces and broadly understood security, using keywords. Monitoring tools enable the simultaneous analysis of publications from traditional media (TV, radio, press) and the Internet (portals, blogs, forums, social media). The service is provided by specialized analytical platforms of the Institute of Media Monitoring, PressService, and NewtonMedia. The presence of defined words and phrases is detected in articles and comments on the Internet, press publications, radio broadcasts, or television programs, both nationwide and regionally. Materials containing passwords intended for monitoring are captured, archived, and made available along with a set of indicators (including reaching, tone, number of interactions), which allows maintaining extensive statistics in the form of charts, summaries, and reports.

Currently, available Internet monitoring solutions provide only simple analyses of Internet users' statements, e.g., based on word counting or analyzing emotional overtones using dictionaries and algorithms from artificial intelligence techniques. Consequently, obtaining an answer to the question that bothers us requires manual analysis of statements by analysts to prepare dedicated descriptive reports. With a massive amount of data, it needs much human work. The next step is to develop technologies that enable full automation of the production of such reports.

In the future, limits on machines' ability to replicate human judgment at speed may be overcome. Some scientists believe that within two decades, a hypothetical future in which machines evolve to have 'Artificial General Intelligence' characterized by the ability to execute the full range of cognitive tasks and, crucially, to understand the context could be realized.

4   Findings

When concluding on the functioning of StratCom, it was necessary, among other things, to analyze the experience of armed conflicts. It was reasonable to conduct observations and draw conclusions from the implementation of StratCom and Info Ops tasks in military contingents' areas of responsibility under the experience use system. The knowledge from the StratCom and Info Ops seminars, workshops, and working groups was significant.

4.1   Training for StratCom and Info Ops

In military universities, the curriculum for education, training, or professional development covers StratCom and Info Ops to a very limited extent.

According to the *Plan of Professional Development at the Ministry of National Defense*, only one Info Ops course is implemented in the Polish Armed Forces. However, no StratCom course is being conducted. The Operational Center of the Ministry of National Defense has developed the assumptions of the StratCom course, which is to be introduced to the course offer of the War Studies University from 2021.

In addition to this course at the Academy of War Arts in Warsaw, as part of the curriculum training at military studies and courses, classes devoted to StratCom and Info Ops are planned. These subjects include Information Operations, Non-kinetic Activities, Social and Media Communication, and StratCom. However, the number of hours is only enough to familiarize oneself with the theory of the problem. In each course in social communication, the Higher Operational and Strategic Course, Postgraduate Operational and Tactical Studies, and Postgraduate Studies in Defense Policy offer 2–4 h devoted to StratCom.

At the Military University of Land Forces in Wrocław, as part of the subject of social communication, cadets are provided with content related to StratCom. In the curriculum of students of the Officer's Study (12 months) for people from the civilian environment, intended for service positions in units of territorial defense forces, the subject of information operations is carried out in 8 h.

At the Air Force University in Dęblin, as part of the study programs and qualification courses for a higher degree, there are subjects familiarizing cadets and students with the issues of Public Affairs in NATO and the Polish Armed Forces, as well as social communication relating, among others, to the general theory of StratCom.

At the Naval Academy in Gdynia, 15 h of Public Affairs are taught on each qualification course for the rank of lieutenant, captain, and lieutenant colonel.

Representatives of the Polish Armed Forces participate in specialist StratCom and Info Ops courses organized mainly at the NATO School in Oberammergau. This training form has covered mainly candidates for positions in NATO structures and key spokesperson positions in the Polish Armed Forces. Annually, only a dozen or so soldiers of the Polish Armed Forces can undergo training in StratCom and Info Ops in foreign centers.

The issue of StratCom has been included in the Polish Armed Forces training system only in recent years. The observations so far prove that this problem has been noticed and is at the stage of the gradual introduction of an increasing number of courses and training at the Polish Armed Forces to develop the management staff's competencies in this area. Education can also be done through practical activities, for example, by issuing StratCom guidelines to generate specific tasks. In turn, it will force the search for solutions, including getting involved in the implementation of StratCom tasks. However, this is hampered by the fact that, although it is imposed by Decision 478, the Ministry's plenipotentiary position for StratCom has not been filled since 2016. There is no such structure at the Armed Forces Branches; only in the Operational Command of the Armed Forces, there is a one-person position. In the General Staff of the Polish Armed Forces, it is the chief specialist in the Secretariat of the Chief of the General Staff, while in the General Command of the Armed Forces, there is no such a position or person.

4.2   Military Equipment

The existing threats and technological progress enforce the need to modernize and acquire new specialist equipment enabling information activities in the modern battlefield conditions. Based on the analysis of threats occurring during armed conflicts, it should be stated that the currently used REA PERKUN type electro-acoustic broadcasters do not meet the safety requirements. The lack of ballistic shields and minimal armor of the equipment units poses a risk of loss of health or life to the personnel in the case of using improvised explosive devices—IEDs by the enemy. Intensive operation of this type of radio stations and the prolonged period of acquiring new ones (REA OMAN) limits the effectiveness of the information impact.

The medium-power electroacoustic broadcasting system (OMAN) is to be mounted on a wheeled light armored vehicle. The broadcasting station is to allow for information activities at the tactical level concerning both soldiers and civilians. It is about carrying out psychological support for general military subunits by recreating electroacoustic messages and special effects of the battlefield. In addition to activities directly related to the military, OMAN is also supposed to provide the opportunity to influence local decision-makers and specific groups when it comes to their will, views, and decision-making processes.

Effective reconnaissance in collecting, processing, collecting, and analyzing data from open-source intelligence (OSINT) is hindered by worn-out mobile devices (e.g., laptops, audio, and video recorders).

The decline in the number of leaflet rocket launchers used to disseminate the products of psychological influence in the future will make it impossible to train the launchers and effectively impact the enemy's forces in direct contact.

4.3   Leadership

When analyzing the afore-mentioned Decision No. 478/MON, it can be concluded that there is awareness of the information environment and StratCom among commanders and their personnel at the strategic and operational level in the Polish Armed Forces (Decision No. 478/MON, 2014). The Decision provisions oblige to take action in this area. On the other hand, finding the answer to whether there is an understanding of the information environment requires conducting research. The studies should also consider lower-level commanders, who are not mentioned in the Decision or other normative documents.

The author has observed a low awareness of the information and StratCom environment and their understanding among lower-level commanders. Not everyone realizes that developing technologies and their universal availability make the information environment complex, heterogeneous, and disordered. Besides, they are also insufficiently aware that information may be distorted by manipulation, bias, prejudices, hidden goals, and both senders and recipients' emotional interpretation.

All military operations have their informational, physical, and cognitive dimensions. They carry a specific communication message. Therefore, it is imperative that higher-level guidelines generate communication activities at all levels. However, there is a lack of coordination of the communication areas already existing in the Ministry of National Defense, including information operations (apart from public diplomacy, social communication, and psychological operations).

5 Conclusions

The multifaceted nature of StratCom (including information operations) makes it a very broad area of analysis and it is impossible to define a ready-made solution that can be implemented immediately. However, the analysis identified several areas where the implementation of corrective solutions is likely to bring positive results.

The following conclusions can be drawn from the observations, analyses of normative documents and the available literature:

1. Due to the current changes in NATO, StratCom is the leading planning element in the field of standardized transfer of information in the command and control system of the Polish Armed Forces in the vertical and horizontal planes. Assumptions adopted at the highest political level (corresponding to the political level of the Alliance) translate in StratCom into the content appropriate to the level of strategic, operational, and tactical command.
2. In NATO, the approach to treating StratCom as a typically advisory and coordinating function for ensuring the compliance of commanders' actions with declarations of political decision-makers is changing. In this sense, StratCom is not an independent function but is responsible for the effectiveness and coherence of existing capabilities. StratCom integrates and coordinates Info Ops, PSYOPS, and MPA, unambiguously ensuring their integration in the planning and executing tasks.
3. The organizational structure of StratCom and its fragmentary nature currently functioning in the Polish Armed Forces, including information operations, does not ensure the implementation of tasks at all command levels. It is purposeful to create a unit coordinating the implemented projects in the field of communication and information activities. Such a solution will enable, among others, the creation of a coherent situational picture in the Ministry of National Defense.
4. The functioning of StratCom in the Ministry of National Defense is regulated by Decision No. 478/MON of December 8, 2014 (MON I, 2014); however, its provisions do not take into account the *Military Policy on StratCom* currently in force in NATO (NATO, 2017).
5. StratCom cells should be equipped with hardware and software supporting the communication process and cooperating with allied systems.
6. The growing importance of StratCom requires education and training in this area, including at the political level superior to the Polish Armed Forces.

The generated conclusions are the basis for the recommendation of corrective actions.

1. It is legitimate to build awareness of commanders and management at all command and organizational levels of their key role in the performance of StratCom tasks. It also applies to the highest state authorities.
2. Solutions related to the integration of StratCom areas should be introduced in coherence with the Alliance solutions.
3. It is appropriate to establish organizational structures appropriate to the needs at a given level of command, which would be responsible for the implementation of StratCom tasks, including Info Ops. Their creation would ensure the proper implementation of the communication process by the Polish Armed Forces.
4. It is necessary to develop a decision of the Minister of National Defense, and doctrinal documents defining the organization and functioning of StratCom in the Ministry of National Defense in the new control and command system of the Polish Armed Forces, considering NATO Military Policy on StratComs (NATO, 2017) and national needs.
5. The implementation of the activities that make up the communication process is possible thanks to the acquisition of IT systems and military equipment enabling effective activities in this area.
6. Finally, it is necessary to verify and adapt the model of education and training of specialist personnel in communication activities in training centers at home and abroad.

6 Summary

The last decade has seen the intensive development of a new concept of active shaping of the information environment, considering the involvement of civil, political, and military factors, and referred to as StratCom (StratCom). StratCom, the essence of which is the redefinition and recombination of processes and structures, and undertakings that have long been present in the public space, especially in its information aspect, is based on the use of the synergy effect achieved from available information capabilities.

The author's observations indicate that there is a lack of StratCom cells responsible for integrating communication effort and coordinating activities regarding communication capabilities, functions, and tools in the daily functioning of the Armed Forces, including during various types of military exercises. That is what happens in the field of psychological, informational, and military social communication operations. A frequent phenomenon is insufficient knowledge in StratCom and the need to use it. The lack of professional expertise affects the insufficient implementation of tasks related to information operations as part of StratCom. One of the critical drawbacks is the frequent lack of staff units responsible for planning, implementing, and coordinating information operations.

Thus, the research hypothesis was confirmed that *the Ministry of National Defense does not have the capacity to implement StratCom (StratCom) at all command levels.*

It is currently not easy to imagine the functioning of state entities without such a comprehensive instrument of broadly understood informational influence, i.e., StratCom. The lack of a comprehensive approach to the organization and functioning of StratCom may result in the inability to use the available communication skills. The conclusions and recommendations in the article should create permanent mechanisms for the exchange of information and views on the organization and functioning of the communication process in the Ministry of National Defense.

While remaining an essential element of the national defense potential, the Armed Forces play an essential role in building and maintaining Poland's strategic potential and should play a key role in shaping the information space in terms of national security. Therefore, it is advisable to implement system solutions in the field of broadly understood communication.

References

ACT. (2010). *NATO Military concept for StratComs*, Allied Command Transformation, 27 July 2010.

Allen, K. (2020). Communicating threat in an era of speed and fetishised technology, *Defence StratComs*, Vol. 8.

Boussios, E. G., & Visvizi, A. (2017). Drones in War: The controversies surrounding the United States expanded use of drones and the European Union's disengagement. *Yearbook of the Institute of East-Central Europe, 15*(2), 123–145.

Campen, A. (1992). *The first information war. The story of communications, computers and intelligence systems in the Persian Gulf War*, Fairfax.

DA. (1996). *FM 100–6, (1996). Information Operations*, Department of the Army, August.

DoD. (2004). *Report of Defense Science Board Task Force on StratCom*, U.S. Department of Defense, 2004.

DoS. (2006). *QDR Execution Roadmap for StratCom*, U.S. Department of State, 2006.

Fredericks, B. E. (1997). Information Warfare at the Crossroads, *Joint Force Quarterly, Summer*.

JCS. (1998). *JP 3–13, Joint Doctrine for Information Operations,* Joint Chiefs of Staff, 9 October 1998.

JCS. (2006). *JP 3–13, Information operations*, Joint Chiefs of Staff, 2006.

JFSC. (2004). *Information operations: Warfare and the hard reality of soft power,* U.S. Joint Forces Staff College, Washington 2004.

Johnsen, F. T., & Chan, K. S. (2019). Military communications and networks. In *IEEE communications magazine*, Vol. 57, no. 8.

Josten, R. J. (2006). *StratCom: Key enabler for elements of national power*, IO Sphere.

Libicki, M. C., *What is information warfare*?, Washington 1995.

MON I. (2014). *Decision No. 478/MON of December 8, 2014 on the StratCom System in the Ministry of National Defense*, MON, 2014.

MON II. (2017). *DD-3.10.1(B)—Psychological operations*, MON, 2017.

MON III. (2017). *DTU-3.10.2—Planning information activities at the tactical level in the land forces*, MON, 2017.

MON IV. (2018). *DD-3.31(A)—Operational masking,* MON, 2018.

MON VI. (2015). *DD-3.6(A)—Radioelectronic warfare,* MON, 2015.

MON VII. (2017). *DD-3.10 (A)—Information operations*, MON, 2017.

NATO. (2007). *MC 422/3 NATO Policy on INFO OPS*, NATO, 2007.

NATO. (2009). *PO (2009) 0141 NATO StratComs Policy*, NATO International Staff, 29 September 2009.

NATO. (2015). *MC 0422/5 NATO Military policy for information operations*, NATO, 2015.

NATO. (2017). *MC 0628, NATO Military Policy on StratComs*, NATO, 2017.

NATO. (2018). *AJP-3.10.2—Allied joint doctrine for OPSEC and deception, NATO*, 2018.

Ranchordás, S. (2019). Nudging citizens through technology in smart cities. *International Review of Law, Computers& Technology, 33*.

Resteigne, D., & Bogaert, S. (2017). Information sharing in contemporary operations: The strength of SOF ties. In I. Goldenberg, J. Soeters, & W. Dean (Eds.), *Information sharing in military operations* (pp. 51–66). Springer International Publishing Switzerland.

SG. (2014). *Instructions for the Organization and Operation of StratCom within the Military Command System*. Sygn. Sztab. Gen. 1674/2014 r.

SHAPE. (2009). *AD 95–2 ACO Directive—ACO StratComs*, Supreme Headquarters Allied Powers Europe, 19 November 2009.

Soeters, J. (2018). *Sociology and military studies: Classical and current foundations*. Routledge.

Stavridis, J. G. (2007). *StratCom and National Security*, JFQ 46, 3rdQuarter.

Szpyra, R. (2003). *Militarne operacje informacyjne*, Warszawa, AON.

USAF. (1998). *Information operations. Air Force Doctrine Document 2–5*, 5, U.S. Air Force, August 1998.

Piotr Szczepański Ph.D Eng., Deputy Dean for Students' Affairs at the Faculty of Security Studies at the General Tadeusz Kościuszko Military University of Land Forces. He gained his professional experience in command positions and as a press spokesman for over a dozen years. Expert in the field of communication management, image creation, and media relations. Co-author of the "Model of Community Promotion" project, which won second prize in Europe in the EUPRIO AWARD, the competition for the best public relations undertaking carried out by specialists from the universities. He cooperates with domestic and foreign universities in the field of image creation. Member of research teams on issues of image creation and the use of IT systems in crisis management. Email: pr.szczepanski@gmail.com

AI and Military Operations' Planning

Ferenc Fazekas⬤

Abstract The key to the success of a military operation rests in its planning. NATO's planning activities are conducted in accordance with the operations planning process (OPP) and implemented by so called operations planning groups (OPG). The pace of technological progress has a dramatic impact on the way the planning process in the military is conducted. In this context, advances in sensor systems and artificial intelligence (AI) play a major role. The leading militaries already use AI-driven tools assisting different steps and phases of planning. The same militaries have also already allocated significant budgets to advance AI research and to develop AI assets, such as automated weapon systems and automated planning assistants. However, the real explosion of AI research and AI assets is yet to pragmatize. From a different perspective, planning-wise AI is frequently viewed as a potential tool to assist human operators. Notably, advances in AI might lead to the introduction of personal planning assistants, endowed with capacities unmatched with those of human operators. These developments weigh on the OPP. The objective of this chapter is to examine how and to what extent.

Keywords NATO · Operations planning · AI · AJP-5 · Comprehensive approach

1 Introduction

The origins of NATO's operations planning process (OPP) can be traced to the second half of the 1990s. At that time, the Guidelines for Operational Planning were developed (NATO, 2006). The first attempt to standardize planning on the doctrinal level was accomplished in 2006 with the ratification of NATO Allied Joint Doctrine for Operational Planning (AJP-5). Despite the promulgation, the doctrine was not accepted until major changes were made, so it was not until 2013 when the AJP-5 was issued as the capstone document for operations planning. The AJP-5 was modified and updated with the lessons learned and with the tested fundamentals of the

F. Fazekas (✉)
University of Public Service, Hungária krt. 9, 1101 Budapest, Hungary
e-mail: Fazekas.Ferenc@uni-nke.hu

© The Author(s), under exclusive license to Springer Nature Switzerland AG 2021
A. Visvizi and M. Bodziany (eds.), *Artificial Intelligence and Its Contexts*, Advanced Sciences and Technologies for Security Applications,
https://doi.org/10.1007/978-3-030-88972-2_6

reigning concept of so-called comprehensive approach. In 2019 the current version of NATO's AJP-5, titled The Planning of Operations, was published. Although the planning activities and fundamentals outlined in the doctrine are applicable to strategic, operational, and tactical levels of war, in this chapter solely the operational implications will be discussed. While the general goals set by the political leadership and the military strategic level tend to be broad and abstract, the tactical objectives must be straightforward and tangible. The tactical commanders on the ground need an accomplishable task and the purpose of it, the reason behind the need of an action. That is, at the operational level strategic expectations are converted into executable actions. This chapter deals mainly with the operational-level activities, as the success or failure of the troops employed depends on the operational plan.

The military organization responsible for conducting the planning process must set up an operational planning group (OPG) from the staff of the headquarters and subordinates. There are already various tools, even AI-driven ones, to aid the OPG in their work. With the increase in the computational speed of the computers these tools are becoming more and more practicable and useful. The current state of AI research suggests that various, advanced AI-driven applications could be introduced in the future to the planning process to facilitate the data processing and to mitigate the risk of potential human errors (McKendrick, 2017; Visvizi & Lytras, 2019a). There is a lot of potential in AI, and there seems to be no limits to its application. If at a certain point in the future the Artificial General Intelligence is achieved, it would change the world as we know it today including the military planning processes Contemporary research highlights that artificial systems may surpass human intelligence and capabilities, dramatically changing the way industry and research works today. In military matters they should pursue objectives that are not aligned with human interests (Visvizi and Lytras, 2019b; Morgan et al, 2020, pp. 24–48; Visvizi et al., 2021).

The utilization of AI in planning is compulsory for every military who wants to keep pace with the innovations. If a military who has access to AI does not use it for military purposes on ethical concerns, there will be the chance of falling behind a potential adversary who does not take the ethics that serious. AI can be used in the field as an intelligence asset or an autonomous weapon system, and can be used in the staff as an aid in planning and assessment. Although there is advanced research in the field of autonomous weapon systems, it is almost certain that the various advanced AI will find their way into the planning teams. Depending on their intelligence level, they could be assistants, functional experts, or even supervisors, but in any case a kind of human-AI relationship develops. Future AI can be a tool, a subordinate or a partner, depending on its intelligence level and on the today unforeseeable legal and ethical developments regarding AI. Considering the likely impact of these advances in AI, the question is **to what extent is it possible to integrate AI into the OPP and, correspondingly, to the work of the OPGs. The objective of this chapter is to do just that. To this end,** the argument in this chapter is structured as follows. Section 2 highlights some important elements of the operational planning process. Building on this, the next section will elaborate on the challenges of operational planning, may they come from the planning environment or from the process itself. Section 4 addresses the role of current AI technologies in the planning process. In the next section potential applications of advanced AI systems are discussed. Discussion and conclusions follow.

2 NATO's Operations Planning Process: An Overview

NATO's operations planning is governed by the Military Committee 0133/4 document, which is the policy behind planning. The actual principles and procedures are set by the AJP-5 (2019) doctrine focusing on the plan development. The AJP-5 clearly defines the planning and its context, while encapsulates the most recent experiences with the comprehensive operations planning. The "how-to" of the process is conducted on the basis of the guidelines set by the NATO Allied Command Operations' Comprehensive Operations Planning Directive (COPD), the most recent version issued in 2021. The conceptual framework of the planning of operations called operational art. The operational art is the employment of forces to attain strategic and/or operational objectives through the design, organization, integration and conduct of strategies, campaigns, major operations and battles (APP-06, 2019, p. 93.). This implies that the operational art deals with the operational design, which is the focal point of the planning process. Operational art is applied by the commanders supported by their staff, and the outcome will be a concept and a plan along which the operations will proceed.

The military operations planning is conducted by an operational planning group (OPG), which will carry out all the tasks defined by the doctrine, the COPD and other functional planning guides. Since it is not a standardized organization, each headquarters has their own approach regarding the planning staff. The OPG usually is not a permanent team of the headquarters apart from its core. Its final composition and manning consolidate during the planning initiation, when the exact nature of the future operation is determined, thus the needed specialists could be involved. The art behind the process is to determine what kind of actions (Ways) required taking into consideration the available capabilities and resources (Means) to achieve the objectives (Ends). This Ends-Ways-Means sequence comes with a great deal of risks (NATO, 2019b), which must be identified and mitigated to have the slightest chance of producing a successful plan.

The planned operations are restricted to a Joint Operations Area, which is a geographical, physical area. The planning team must be aware of the situation and proceedings of this area through the entire process, but the planners and their commander must be aware of the wider picture also, namely what happens in the Operating Environment. The Operating Environment (formerly NATO and current US: Operational Environment) is a composite of the conditions, circumstances, and influences that affect the employment of capabilities and bear on the decisions of the commander (NATO, 2019a, p. 93). These conditions, circumstances and influences vary from operation to operation, and they also evolve during each operation. The actions of the military may trigger undesired changes and move the conditions towards an unacceptable state. The plan produced on presumed facts and assumptions regarding the operational environment may change completely by the start of the operation itself, which requires the constant adjusting and revising of the plan, putting pressure on the OPG. The operational-level planning starts when the strategic level issues its directives, guidance and analyses which form the base of the process. At first the initiating directive and later the guidance sets the boundaries for the planning, providing the strategical context.

The essential element of planning is the information. If information is scarce or not valid, the entire process is doomed. The intelligence preparation of the operating environment is a never-ending process, information that is valid at the beginning of the planning could have been changed by time. The intelligence analysis is an ongoing element of the entirety of the planning process, which is demanding for the human resource.

The NATO OPP uses a holistic system-of-systems approach for breaking the complex operating environment into evaluable pieces that the plan will build upon. Framing the problem is a key element, during this several analyses will happen, including the actor, factor and the center of gravity analyses. These analyses result in a system view of the operating environment as we see it from military perspective. Once all analyses are complete the course of action development starts depending on the commander's guidance and the data processed so far. The planning steps regarding courses of action uses every product of the previous steps to find a solution between the identified boundaries. There are a couple of adversary courses of actions devised reflecting what the adversary can possibly do and what is the worst-case scenario for the friendly forces. There are also several courses of action for the own forces, each of which will be tested against every adversary. As a result of these tests and the analyses of the courses of action the commander will be able to select the most viable one for the specific operation. From this point on the remaining planning activities are the elaboration and refinement of the concept devised during the course of action development.

The importance of the current method regarding the usage of the OPGs comes from the fact that the sequence of planning activities are conducted by the same team, each and every member has the same understanding of the situation, knowing the problem from its very root to the proposed and planned solution. This team with its deep knowledge of the unfolding situation has a better chance to react to any relevant change of the operational environment.

3   New Challenges in the Military Planning

The current operations planning process is the best available, but it cannot be suitable and ideal for all situations. The successful methods applied in the past might not be as effective today in a constantly changing environment. With this constant change new challenges emerge for the planners, while our existing military planning process still contains some inherent challenges. The first of the inherent challenges is rooted in the doctrine and the directive itself. As the COPD and the AJP-5 are 6 years apart, these documents may contain some minor inconsistencies. This is not a critical issue, but if there are possible different understandings of the same entity then it is a potential cause of failure. The operational art, as the framework of military operations, is called "art". It is not an exact science and, as the NATO doctrine emphasizes, it cannot be learned from books, one needs intuition, experience and skill (NATO, 2019b, p. 2–3). (. In other words, as in art, so in the military, one may have the required attributes to be an artist or not. The leader, who is responsible and their planners, who are conducting the design and organizing of the operation have to reach back to their previous experiences from similar situations. Sometimes they just have to rely on their instincts, imagination, and making the best guess (March, 1994). Reliance on previous experiences is the second inherent challenge of the planning, and this is also where the third inherent challenge lies: there is no guarantee, that the available specialists have the knowledge needed to solve the problem ahead, and there is no guarantee that the needed specialists can be gathered on time, if ever. Even if the required number of personnel is ready with all the required capabilities, the fourth inherent challenge comes to play, which is the human factor: the chance of committing errors due to overload, exhaustion, or any other issues. To mitigate this most planning activities are conducted in teamwork, with different sections and syndicates discussing and cross-checking the developed products, but the chance of error still cannot be ruled out. In situations never before seen the planners still have to rely on the best guess, which may fail the test of execution.

Emerging novel situations and factors have the potential to change the planning process, either all at once or in a piecemeal fashion. Current NATO approved operations planning procedures are using a comprehensive approach, which involves the coordinated application of the diplomatic, information, military and economic instruments of national power (NATO, 2019b). Earlier, during the Cold War era, even until the invasion of Afghanistan and Iraq in 2001 and 2003 respectively, the governing idea in military planning was so simple: it is the enemy against the friendly forces, red vs. blue. Operations were planned with little to no regard for the local population or the interests of religious groups, the only thing counted was the defeat of the enemy fighting force. Lessons learned in Iraq and Afghanistan changed this view: insurgent troops fighting each other as well as the invaders, tribal communities acting along motives incomprehensible for soldiers made clear that the nature of warfare changed. Military means alone were not enough to meet the complex security challenges. To execute effective crisis management the comprehensive approach was developed. Achieving comprehensiveness is a challenge alone itself. High level of cooperation and coordination is needed along these instruments of power, while their inner structure and functioning are not comparable.

From the very nature of the military, it is evident that capabilities and resources are limited in a military organization, but new challenges may need new capabilities. In NATO, a capability is defined as the ability to create an effect through the employment of an integrated set of aspects categorized as doctrine, organization, training, materiel, leadership development, personnel, facilities, and interoperability (NATO, 2019a, p.23). All of these elements has to be addressed in building new capabilities, which takes a lot of time after the need for them identified. In novel situations the challenge is to solve the problem on hand with existing capabilities.

The operational environment in which the operations will take place is a complex environment in constant change. The challenge is to keep the pace with these changes, and to conduct the needed plan revisions and updates in a timely manner. The world in which NATO militaries must stand and do their job is not the same world that these militaries were designed for. NATO published its strategic concept titled "Active Engagement, Modern Defence" in 2010. The operational environment underwent significant changes since then. The US Army Training and Doctrine Command developed a pamphlet in 2019 which describes the possible conditions the US Army is about to face in the following 30 years: these are likely very close to what NATO forces will need to face. The Operational Environment envisioned is affected by the internet of things and social media, the changing demographics, the competition for resources and geopolitical challenges. Major challenges will be the rapid societal change triggered by technological and scientific advances, and also the potential adversaries' ability to modernize and adjust their capabilities to the changing Operational Environment. The aforementioned factors may trigger and fuel new conflicts, mass migration and governmental changes (TRADOC, 2019). To be able to operate in different roles, as warfighters or as peacekeepers is a major challenge for military forces, and requires different mindset from the planning phase on.

4 Application of AI in Planning

An almost constant technical revolution multiplied the information attainable to the analysts and planners to the point, where even the smartest human brain can no longer process all of them in the short time usually available. OPGs already have a range of tools to aid their work, but using AI to filter the information, predict events and reactions could be a game-changer for the OPP. The challenges to solve a problem with limited resources, in a limited timeframe and in a highly volatile operational environment in close cooperation with other institutions require novel technologies. These improvements can open new horizons and highlight opportunities which may be overlooked by a human-only OPG. As demonstrated in the preceding sections the operations planning process is massively information driven, and encompasses a series of recurring analyses. The apparent points where AI can be employed are these analyses, since many of them have a logical reasoning.

AI in its current state is proficient in solving specific real-world problems, it can engage deep learning to find patterns using huge training datasets. The problem is that in military operations planning there are only a handful of specific problems, the problems vary from situation to situation. That is also affect the deep-learning approach: an algorithm trained in a specific task environment may not be successful in a different operational environment of a different theater of war. An algorithm designed for peace support operations may have limitations when it is to be used in conventional war because of the different fundamental principles of the two types of operations.

Engaging AI in planning is not a novel idea, the US Army heavily utilizes its AI planning agents since the 1990s. The Dynamic Analysis and Replanning Tool (DART) made invaluable services prior the First Persian Gulf War in 1991 (Lopez et al., 2004). The Joint Assistant for Development and Execution (JADE) is a tool for assisting force deployment planning in short time. Although these systems are highly effective, their product is up to the standards and using different AI algorithms, they still require the operator to interact with. As the developers of JADE admitted they overestimated the deployment planning skill level of the typical JADE user (Mulvehill et al., 2001, p. 14). Nevertheless, these tools are primarily for strategic-level solutions. These solutions still have to be translated to something comprehensible and executable for tactical-level echelons.

NATO has its own toolset for operations planning called the Tools for Operations Planning Functional Area Services (TOPFAS). The system itself consists of several integrated tools, which are all contributing to and working from the same database thus enabling the achievement of a common operational picture along the physically distributed, network-connected workstations. These tools help visualizing the systems of the operational environment, the planned operation and later on the effectiveness of the execution (Tamai, 2010). While they are useful and save time for other activities, they are operator-driven and not able to produce new information the way as expected from an AI.

There are many potential fields for application of AI during the OPP as shown in Fig. 1. Every single step, which is more of a logical procedure than a complex problem solving challenge could be left to the AI. Analyzing raw intelligence data is just one among these, while it is the most tiresome one, the produced information is

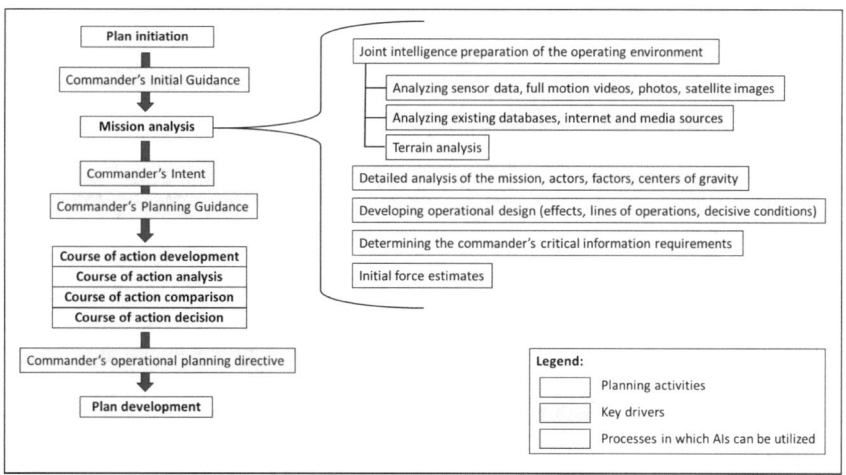

Fig. 1 Outline of the OPP *Source* created by the author, information based on NATO, 2019b

the fundament on which the entire process depends. The joint information preparation of operating environment contains a lot of data processing to produce information needed for the planning. During the twenty-first century sensor technologies and data sources have been revolutionized, each of them producing exponentially more data than their counterparts from the Cold War era. All new sources provide data in abundance, such as satellite imagery data, full motion videos and reconnaissance footages from the air, military intelligence and reconnaissance troops on the ground, not to mention the data from the internet, social media, online media and even private correspondences. Whatever the source of the data is it needs processing before it can be used by the OPG (Fig. 1).

When the OPP is ongoing any changes in the assumptions or facts may alter the entire course of the planning, thus implementing AI to produce, to compare or even to analyze the information is essential to retain the initiative against the adversaries. All the routine processes, which require humans to follow certain steps should be automated. This automation can be achieved either by traditional computer applications or by AI in order to reduce data process time. AI outperforms humans in tasks require great amount of concentration or computation. AI's main role is to reduce the cognitive overload of the limited number of planners.

The human and AI relations are today restricted to feeding the AI with valid input data and using the AI-produced output information. The interactions are usually restricted to a user interface through which the AI can have the inputs or instructions it requires. There is a potential to change this, for speech recognition and natural language processing is a promisingly developing field of the civilian AI research. Independent and government-founded companies and organizations are working on achieving new breakthroughs in the AI fields, and these progresses must be monitored and the suitable ones have to be adopted by the OPGs as soon as possible to improve effectiveness. Being more effective save lives either in peace support operations or in conventional engagements.

5   The Future of AI in Planning

As computer technology advances forward new possibilities will be born. The universal quantum computer built by Google AI in 2019 solved a problem which no existing supercomputers could have solved in a reasonable timeframe, even with massive optimizations it would take days for a supercomputer compared to minutes of the quantum machine (Arute et al., 2019). Chinese scientists successfully experimented with a different quantum approach, although their machine is not yet programmable, they claim it solved a computational problem in 200 s which would take 2.6 billion years for an existing supercomputer (Zhong et al., 2020). It can be assumed that universal quantum computers become more practical and accessible with time, and the AI based on quantum computers will be more efficient than their counterparts developed on traditional computers. Even the mere presence of cutting-edge quantum-powered computers will provide superiority in the future operations against those, who does not have it yet, or has it only in an obsolete form.

The ultimate challenge of AI development is to produce an AI which can act and think as human. From a planner point of view that is definitely not desirable: since the human intellectual mechanisms are flawed, we need something more precise and faster than human thinking. The human decisions may base on preconceptions, may be bounded by cognitive biases, governed by heuristics, any of which may speed up decision-making, but they carry the chance of failure. A well-functioning AI should not do these, or, if it has to, it should provide a warning, or a percentage of probable success, highlighting the unknown elements. But technology is still far away from these kind of AI realizations. Fully functioning operational planning support AI needs to be specific constructs which are able to appreciate the many aspects of an operating environment. There is a possibility, that in the distant future one AI can solve all the problems that the OPP encounters, but until then many different functional AI should be developed and used simultaneously, all of them contributing to the same shared knowledge base.

The most crucial field of AI application is the joint intelligence preparation of operating environment. Image processing is a field in which AI already made a good use, and there is room for improvement. For instance, recognizing troop movements and firing positions from satellite images or other reconnaissance products, detecting staging areas, forecasting the intentions of the adversary forces are all essential to plan and direct the operations. Conducting terrain analysis depending on real-time data instead of printed maps and orthophotos can also be an advantage over the enemy. Using AI to these different tasks and then combining their products to make a virtual, 3D presentation of the joint operations area gives an important asset to the planners and commanders. This invaluable asset is the ability to visualize the physical area in which they have to operate, with all its features, possible force locations, avenues of approach, restricted terrain, obstacles and virtually everything they want to see. This depiction of the operating area, a kind of common operating picture interface can integrate all the actions and activities in different layers, which all other military or non-military entities may execute, thus taking the cooperation and coordination to a whole new level, contributing significantly to achieving real comprehensiveness.

Novel challenges may require novel use of existing capabilities. There lays the importance of human creativeness, as it requires the art of constructing new ways for existing means. AI can be of assistance in this case, too. Based on pattern-recognizing the deep learning AI algorithms may be able to highlight past similarities to the current situation, and may provide examples from previous operations. These examples may serve as a base of thinking for planners. An advanced AI could go as far as making up courses of actions depending on the situation of the own and adversary forces, taking into consideration all the information available in the common operational picture interface. It may not be possible for AI to produce executable courses of action, it may just assist the human planners in developing one. The real-world issues of testing and analyzing courses of actions, which usually takes very long time can be shortened considerably by the application of AI that can run thorough simulations to test the validity and feasibility of each course of action in a matter of minutes, highlighting potential problems.

The challenge of diversity and versatility of the complex real-world situations require methods that are able to react in a very short notice to the emerging changes and difficulties. The AI applications mentioned above, i.e. the data-processing, operational picture updating and course of action developing tools are designed for exactly this purpose: saving time. Time is a crucial factor that can save lives. Saving other resources is important also. Speeding up the planning and plan updating with intelligent tools can be a decisive element in future warfare.

Human memory has its limits, but an AI's memory is mostly hardware-dependent and can be expanded. While a human planner may have experience and obscure memories from a limited number of operations or campaigns, plan-aiding AI assets can have the memory of other AI also. AI can learn from the experience of other AI, and they will never forget or overlook important factors, if they applied and programmed correctly. The problem of different mindset required for conventional military and peace support operations can also be mitigated by applying AI that can differentiate between the methods, and will be able to highlight inappropriate solutions.

With the application of AI planner systems the inherent challenges of operations planning can also be mitigated. If the AI has the shared memories of all other AI ever applied to a specified field as described above, it can provide such a great deal of previous experiences that no other human may possess. It is also possible and desirable to provide AI for each functional areas and special staff positions which may analyze options, offer solutions or even just hints to their human counterparts. This human-AI interaction can open whole new approaches to operations planning which are not conceivable today. This kind of AI assistance not only raises the experience-level of the OPG, but also reduces the chance of error inherent in human cognitive processes. As technology progresses more and more applications can be found for AI in military planning, as well as in other fields of our life. Civilian AI applications are spreading, which most likely result in their increased military application soon. Combined with other possible advances in computer technology the prospect of portable AI assistants is not just an unreachable science-fiction cliché anymore.

6 Conclusions

The current NATO operations planning process uses a holistic approach to find solutions to the challenges of the complex real-world situations. These solutions are made up of information processing data gathered by reconnaissance and intelligence assets as well as produced by analyzation of different aspects of the operational environment. The resulting information pool is used to design the possible courses of action, upon which the concept of operations relies. This process is demanding, a lot of variables and attributes complicates the rational decision-making. The implied analyzation methods are designed to simplify the complex environment allowing the human mind to comprehend it. Paying attention to every detail is demanding and as the situations are getting more complex, it becomes almost impossible. The operations planning groups are using teamwork procedures to verify each other's findings, such as the deductions and conclusions drawn during the mission analysis. This way they are able to rule out misleading information and avoid misdirected actions. Thorough planning of a NATO crisis response operation may span from weeks to months, and this is a resource and time-demanding process. Intensive AI usage can reduce the pressure put on human operation planners and can speed up decisively the process contributing to retaining the initiative.

Various AI tools are used present days to facilitate the military planning processes. Most of these have little use on the operational level of war, since it is a unique field, where the military science and the operational art meets: due to its artistic nature not every aspect of it can be automatized. There are useful tools to make operational-level planning work easier, there are a lot of possibilities to introduce new AI-driven tools and methods to the process. Judging on the present state of AI research there will not be any AI capable of executing an operational-level planning process on its own. If technologies will be available to set up a human-AI mixed planning team as envisioned in Chap. 5, potentially every functional expert may have their own AI assistant. These tools will have the experience of decades of previous operations and exercises, and this would give an upper hand against an adversary of ordinary human cognitive capabilities. Though one must be aware that the flaws and faults of the previous operations will also be there in the databases. An AI today is as intelligent as its programmer, and predictably this will not change in the future. If an algorithm has built-in flaws then the use of AI is a risk rather than an advantage. Achieving a prudent AI is not enough: AI has to be failsafe and protected from adversarial cyber activities.

For the time being the viability of an independent AI planner system capable of human-level performance remains low, so AI engaged in planning will be most likely AI assistants. Either way, operational planners must not build a complete reliance on AI since there may be situations when electrical devices are knocked out of play. In these cases they have to return to the traditional planning activities, without the use of computers and other intelligent devices. This is like a worst-case scenario, but however low its chance is it may happen.

The emerging possibility of applying neural implants which are designed to facilitate human–computer dialogue is also a factor to be considered. If one will be able to communicate with their AI assistant in an instant via neural activities, it would speed up things even more, multiplying the possibilities the AI currently have. This achievement is still far away, but its possible impact may not be underestimated.

Operations planning definitely need the AI: assisting in the current time-consuming and human resource-demanding processes are a logical field of its application, there are already some AI-driven assets in use, with more to come. The final level, where AI will be integrated in an operations planning group remains unknown and depends on the future development of hardware and software technologies. The prospects are bright, with the current level of technology there is space for AI tools.

Well-trained and experienced human planners augmented with sophisticated AI planning tools can shape the future's military operations. If there was an AI sometime in the future that is capable of operational-level planning individually independent of human operators, it would raise ethical questions. May an AI decide the application of human resources without restraints? Will there be any human on the future battlefields, or AI-only weapon systems will engage each other on the plans of AI planner systems? As long as there are humans involved, either as combatants or as civilian population in the joint operating area, there must be human supervision of everything done by an AI.

So, to what extent is it possible to integrate AI into the operational planning group? AI can be the workhorse of the OPG, but regardless their intelligence level they must be secondary members. The possible human-AI relationship issues and rules for building an effective team are already discussed by researchers (Joe et al, 2014), but until the emergence of advanced AI this ethical question can be suspended in the context of OPGs. While the decision remains in human hands the AI will be inseparable part of the OPGs in the near future. AI and their products must be monitored and checked, for a complete trust requires the conviction that the AI functions as its designer wanted it to do so, and no external adversary infiltrated into its systems. In our age of the intensifying cyber competition this possibility may never be ruled out completely.

Acknowledgements The author wishes to thank Dr. Anna Visvizi for the comments, ideas and corrections that facilitated the realization of this chapter, and to Dr. Zoltán Jobbágy for drawing the author's attention to this opportunity.

Disclaimer: It is important to stress that the statements and conclusions are the author's own and do not reflect official views of either the NATO or other national militaries.

References

Arute, F., Arya, K., Babbush, R., et al. (2019). Quantum supremacy using a programmable superconducting processor. *Nature, 574*, 505–510.

Joe, J. C. et al. (2014). Identifying requirements for effective human-automation teamwork. Probabilistic Safety Assessment and Management PSAM 12, June 2014, Honolulu, Hawaii. https://inl digitallibrary.inl.gov/sites/sti/sti/6101795.pdf

Lopez, A. M., Comello, J. J., & Cleckner, W. H. (2004). Machines, the military, and strategic thought. *Military Review, 84*(5), 71–76.

March, J. G. (1994). *A primer on decision making: How decisions happen*. Free Press.

Mulvehill, A. M., Hyde C. & Rager D. (2001). *Joint Assistant for Deployment and Execution (JADE)*. Air Force Research Laboratory.

McKendrick, K. (2017). The application of artificial intelligence in operations planning. NATO. https://www.sto.nato.int/publications/STO%20Meeting%20Proceedings/STO-MP-SAS-OCS-ORA-2017/MP-SAS-OCS-ORA-2017-02-1.pdf

Morgan, F. E., Boudreaux, B., & Lohn, A. J. et al. (2020). Military applications of artificial intelligence: Ethical concerns in an uncertain world. RAND Corporation. https://www.rand.org/pubs/research_reports/RR3139-1.html

NATO (2006) AJP-5 (2006) NATO allied joint doctrine for the planning of operations ratification draft, NATO Standardization Office.

NATO (2019a) AAP-06 NATO glossary of terms and definitions (English and French), NATO Standardization Office.

NATO (2019b) AJP-5 (2019) NATO allied joint doctrine for the planning of operations, NATO Standardization Office.

Tamai, S. (2010). Tools for operational planning functional area service: what is this? In *NRD-C ITA Magazine,* (14), 20–22.

TRADOC (2019) Pamphlet 525–92 the operational environment and the changing character of warfare, Department of The Army.

Visvizi, A., & Lytras, M. D. (Eds) (2019a) Research and innovation forum 2019. Technology, innovation, education, and their social impact. Springer. https://www.springer.com/gp/book/978-3-030-30808-7 ISBN 978–3–030–30809–4

Visvizi, A., & Lytras, M. D. (Eds) (2019b) Politics and technology in the post-truth era. Emerald Publishing. https://books.emeraldinsight.com/page/detail/Politics-and-Technology-in-the-PostTruth-Era/?K=9781787569843. ISBN: 9781787569843

Visvizi, A., Lytras, M. D., & Aljohani, N. R. (Eds) (2021) Research and innovation forum 2020: Disruptive technologies in times of change. Springer. https://doi.org/10.1007/978-3-030-62066-0, https://www.springer.com/gp/book/978-3-030-62065-3

Zhong, H. S, Wang, H., & Deng, Y. H. et al. (2020). Quantum computational advantage using photons. *Science.* https://science.sciencemag.org/content/early/2020/12/02/science.abe8770

Ferenc Fazekas graduated in 2003 as infantry officer, and served 13 years in a mechanized infantry brigade as platoon leader, company commander and brigade operations officer. During this time he made 5 tours of duty to various countries, including Afghanistan and Iraq. In 2016 he became a staff officer in the Hungarian Joint Force Command J5 (Plans and Policy) branch, charged with military operations planning. After getting master's degree in Military leadership he joined the Ludovika University of Public Service as assistant lecturer, and was assigned to the Faculty of Military Strategy. He currently works on his doctoral thesis, which builds on the experience gained while working with operational level problems using the NATO's comprehensive approach.

AI-Supported Decision-Making Process in Multidomain Military Operations

Wojciech Horyń, **Marcin Bielewicz**, and **Adam Joks**

Abstract Modern military operations are conducted in a dynamically changing environment, which necessitates that huge volumes of data are analyzed instantaneously. Due to advances in information and communication technology (ICT), decision-makers can be supported in the process of accessing and making use of the vast amounts of data and information. However, due to the instantly available large amounts of data, they are not able to make optimal choices. Technology once again helps providing several solutions to aid decision-makers in making timely and well thought decisions. The operational environment is filled with various devices able to collect, analyze and deliver data in a split of a second. They are driven by powerful AI engines, use modern simulation software, and use machine learning algorithms. It seems that the commanders and their staffs receive the right and ultimate solutions. However, all that technology is devoid of a social dimension, and their cold analysis is deprived of objectivity and humanity that are essential in making decisions in the modern operations. The authors will investigate the usefulness and risks of this type of decision support in light of modern and demanding military operations.

Keywords Artificial intelligence · Simulation · Data farming · Machine learning · Decision making process

W. Horyń
Faculty of Security Sciences at the Military University of Land Forces, University of Business, Wrocław, Poland
e-mail: wojciech.horyn@awl.edu.pl

M. Bielewicz (✉)
International Journal of Security Studies of the University of North Georgia, Dahlonega, United States

A. Joks
Faculty of Political Science and International Studies at the University of Warsaw, Warsaw, Poland

1 Introduction

Decision making in crisis situations is a complex task that requires solution-seeking ability under pressure with changing factors and often needs action of other people. A process that requires actions to be optimized simultaneously leads to solving multi-objective problem, which further requires use of more resources in seeking efficient solutions (Gonzales et al., 2011, p. 1153). Artificial intelligence (AI) offers humanity a hope, especially in complex tasks, primarily mitigating a human error. Although the evolution of artificial intelligence could be mapped to 17th century and Leibnitz's idea of understanding principle of reasoning, designing its principles and build a system corresponding with the design, it was not until later, in 19th century, when AI started to be seen as a complex adaptive system that would lead to more efficient tools in complex decision-making process (Spector, 2006, p. 1251). The dilemma arises in connection to human independent thinking, social context, and individual expression while making decision.

The chapter examines the possibility of using AI-powered tools in the military decision-making process. In this context, the caveats of uncritical use of AI are discussed. Against this backdrop the objective of this chapter is to address the question to what extent do technology analytical tools affect the decision-making process and to what extent can human-being be replaced with artificial intelligence in decision-making?

For that purpose, the authors investigate the capabilities of SWORD simulation and analytical tools as a case study, and discuss the practical and social aspects of using technology in the support of decision makers.

The chapter is structured as follows. The first part presents the decision-making process and describes the specificity of it in the military organization. It is argued that military leaders' decision-making process has far-reaching consequences. Among other things, wrong decisions may result in the destruction of infrastructure and lead to the loss of life among soldiers and civilians. The decision-making process is characterized by a large amount of information to be analyzed in a relatively short time. Therefore, the support of modern technologies is increasingly used. Similarly, the Battle Management System (BMS) is presented as the practical application of AI support for military commanders. Additionally, they indicate that using the support of artificial intelligence in the decision-making process, various simulations: live, virtual, constructive (LVC), should be considered, depending on the command level. The second part of the article describes the features of the SWORD system as the case study, pointing to its practical application in the decision-making process by military leaders. The third part is a discussion on the possible use cases of emerging and disruptive technologies (EDT) in a decision-making process. The authors indicate that technological development is very advanced, artificial intelligence is no longer just a fantasy, but is becoming a reality. The use of neural networks or machine learning in artificial intelligence makes it closer to a human brain.

2   The Decision-Making Process

The right choice of a decision depends on a properly executed decision-making process. The goal of decision-making process is to determine the best alternative out of possible variants. This process is based on both, logical and algorithmic information selections (Scibiorek, 2018; Triantaphyllou et al., 1998).

The decision-making process encompasses many areas, including economy, management, business, crisis management and the functioning of the armed forces. When analyzing the implementation of tasks by the armed forces, the decision-making process is its indispensable element. American political and military science consider the decision-making process as planning methodology that integrates the activities of the commander, staff, subordinate headquarters, and other partners to understand the situation and mission, develop and compare courses of action (COAs), decide on a specific COA that best accomplishes the mission, and produce an operation plan or an order for execution. Decision-making process helps leaders apply thoroughness, clarity, sound judgment, logic, and professional knowledge to understand situations, develop options to solve problems, and reach decisions, therefore such an approach to the decision-making process is universal and includes a practical reference to actions taken also beyond wax structures (CfALL, 2015, p. 7). The issues concerning decision-making process have been widely discussed in the literature (Agrawal, 2020; Clemen, 2001; Hamilton & Herwig, 2004; Toner, 2005; Schraagen & van de Ven, 2008; Vroom, 1973). Still, there is no consensus as to the phases/stages of the decision-making process. The literature (Scibiorek, 2003; Wisniewski & Falecki, 2017) suggests a decision-making model that can be divided into following phases: identification of the decision problem, searching for information (knowledge about the analyzed decision topic), decisions (i.e. choosing a solution from many variants) and evaluation of the decision made.

An important element in the decision-making process is collecting and then analyzing information and developing an appropriate variant of action. When transferring the decision-making process to the functioning of soldiers in military operations, information and the decision-making process are of fundamental importance. The multitude of information provided info and the minimized time to decide, create new possibilities in the use of artificial intelligence. This approach supports the activities of soldiers at the initial stage and might lead to their removal (soldiers as a human factor) at the stage of implementation of advanced disruptive technologies.

The complexity of the modern world dynamics makes any military operation, regardless of their breadth and width (strategic, operational or tactical), type (full-scale, stability or peace) or composition (allied, coalition or national), single or multi-domain (air, land, sea, cyber, space) difficult to organize and conduct. The necessity to include into decision-making process various traditionally non-military factors, such as media, international governmental organizations (IGO), non-governmental organization (NGO), civilian population, social media, *etc.* adds more complexity to the operations for decision makers on all levels. To cope with many preexisting factors caused by certain domains and their possible impact on operation, the commanders must employ or outsource increasing number of specialists.

The amount of data collected requires adequate time for processing. To speed up the process, there are various systems and protocols. They assist staff and experts/analysts in categorizing, analysis and building the picture that helps visualizing the possible impact on the outcome of the operation or its phases. The more complex environment the more factors affecting the decision-making.

Battle Management System (BMS) can create common operational picture, as long as either it is a single version (producer) system or there are interconnected systems of BMS systems using the same data format or they communicate with the use of bridge or protocol that translates data back and forth. Otherwise, it is impossible to use data of various types, versions and releases. This could lead to not full representation of situation and false situational awareness of the commander making decision. This could result in such dramatic outcomes as: unintentional deaths among non-combatants and destruction of internationally protected sites.

It should be stressed that we are talking here about the easier part of it—a presentation of the past and current situation. When we take into consideration a decision-making support, we would like to have a little bit of the future and possible outcomes of the decisions we would have taken. It is not however what some could call "fortune telling" but rather the course of action validation that, with the support of the technology, presents possible results of each CoA before it is implemented. It helps estimation of benefits and costs and then selection of the particular CoA.

Very useful tools for this purpose are the simulation system. They are the software systems that create environment giving opportunity for multiple user's interactions in a real-time from various locations (Hodson, 2009). Simulation systems use very sophisticated algorithms to calculate input data, play scenarios and visualize the effects of the decisions made by commanders (decision-makers), and input by the operators (responsible for tasking the units and elements they control).

There are generally three levels of simulation systems: live, virtual, and constructive (LVC):

- live simulations are those where real people operate real instrumented systems, like weapon operator (gunner) during the field training, where effects of the fire are simulated;
- virtual simulations are when participants operate a simulated equipment in a simulated environment, *e.g.* tank gunner in a tank simulator;
- in constructive simulations, both operators (humans) and equipment are simulated in a simulated environment (Simulation: Live, Virtual and Constructive).

Among the above mentioned three types of simulation systems, the most useful for decision-making support are constructive simulations. They allow to play various scenarios with large number of entities on vast terrains. There are of course constraints in use, mainly due to quite simple automation modules they employ. This has both advantages and disadvantages, which originate in the philosophy of their creation.

The constructive simulation systems were developed to support command and staff training. Therefore, by default, there was requirement of operators for each operational unit that was represented in the scenario. The level of such unit (*e.g.* company, battalion, regiment/brigade, division, corps) depends on the particular software category and its dedication (tactical, operational or strategic level). Hence, the reliance on real operators' control over each unit does not require any sophisticated automation tools embedded in the simulation software. The advantage of such solution was to have real human action and behavior while controlling units, which required simple automation modes for lower echelon units and entities. However, this made experimentation and course of action validation difficult and costly. Additionally, the force reduction caused mainly reduction of support and auxiliary personnel. Hence, most simulation centers lost their permanent constructive simulation operators, which in turn required the training unit to provide the operators. It is not a big issue during the training in peace time, when commanders might dispatch soldiers from not participating units to this role. In the operation the unit personnel has its specific role to play, and their absence from it might impact the outcome of the battle.

Taking the above mentioned factors, the industry, on the end-user's interest, started to investigate possibility and experiment with artificial intelligence. It was not new that simulations used sophisticated behavior algorithms, but it was limited mainly to the virtual simulations, which were basically robust versions of commercial computer games. A good example of such approach is the commercial game the *Armed Assault* by Bohemia Interactive company and the *Virtual BattleSpace* tactical simulation software series (VBS2, VBS3, VBS4) by its partner Bohemia Interactive Simulations.

Following this example, some constructive simulation producers, such as MASA Group, released their constructive simulation software that consists of AI engine. This move accepted by some of the European countries, created larger flexibility for experimentation and course of action validation. It is no longer necessary to employ lots of operators on many workstations. It is enough to create the scenario and specify the conditions, and run the exercise, course of action or experiment on single computer with single operator. Software such as MASA SWORD constructive simulation allows aggregation and de-aggregation on the fly, changing the speed of simulation clock, and dynamic scripting depending of the requirements. It saves time, money and space, because it could be run on a laptop in the field command post without robust network infrastructure.

3 SWORD (Masa Group) Analytical Tool Case Study

The purpose of a SWORD Analysis, the SWORD constructive simulation add-on, is to be able to test and analyze, compare and visualize, as well as present outcomes of courses of actions. It allows more comprehensive analysis of variants during the Military Decision-Making Process (MDMP). It is achieved by flexibility of this solution that by design is flexible providing scalability that helps adjust it to a need, whether it is operational analysis, education, decision support, planning assistance, concept development, experimentation or computer assisted exercise (mainly as the support for after action review of the CAX).

Classical simulations using certain aggregation level require number of operators control entities (*i.e.* units, elements, objects) to run command post/staff and emergency management exercises. The operators are not a training audience and are limited to receive instructions and reporting the outcome (situation change). The requirement for operators creates human and financial burden difficult to contain, if the user intends to conduct an increasing number of training and/or tests per year. With the use of artificial intelligence and automated modes SWORD provides the user with flexibility in this regard.

Simulations need to be highly usable if they are to be efficient, therefore the content of the simulation itself needs to be customizable and adaptable to the specific operational context of the simulated situation (military operation, emergency response operation, peace support operation). It must allow customization of equipment, weapon systems, sensors or any other doctrine-related element, thus to be easily modified and adapted by the end-user. Another advantage of this simulation software is the ability to either train an operator quickly or to have operators self-trained, with the prepared robust help library in pdf format and interwoven with the simulation. Moreover, the use of the MASA DirectAI engine (the producer's proprietary Artificial Intelligence technology) to drive the synthetic forces as non-player-characters (*e.g.* subordinate units, support units, enemy and neutral entities, *etc.*) allows the operator to focus on command-related topics, rather than divert his attention to the functional low-value operation of each of the different units or subsystems. It empowers simulated entities with doctrine-compliant behaviors. It ensures maximum realism while requiring fewer human controllers, and fully focused on the core tasks. Moreover, the behavior models can be modified by the user (the system administrator and exercise director staffs), which allows adjustment to represent differences in behavior of various entities from different military cultures (*e.g.* various doctrines) and local population customs.

SWORD constructive simulation is able to simulate the behavior of up to 10,000 units (this number depends on the level of aggregation) on the terrain up to 500 × 1000 km, on the ground, sea and in the air. It allows to extend these capabilities by connecting via standards like DIS, HLA or BML, a number of servers, each hosting part of the scenario. In addition, the native support allows connection of SWORD with larger community of simulation and information systems (e.g. command and control systems).

The system was designed also with consideration on the irregular (asymmetric, hybrid) warfare concept. As a result the simulation include various modes rather than provides "kinetic effect-based" simulation. The DirectAI engine uses a library of behaviors that have been written, modeled and validated, to specifically support activities of refugees, militias and terrorist. These behaviors are autonomously driven, that means the virtual entities, either a human or an object- reacts and adapts its behavior to changes in their synthetic environment without the need of any operator's intervention. New behavior libraries can be added, and new missions can be specified and modified over time without affecting the core simulation. Specific modules supporting emergency management scenarios and behaviors models extend software capabilities even further, *e.g.* the software includes a population density macro-model that allows for a low overhead population or urban environments in coordination with the synthetic force elements operating in virtual world. Modeling of fire, flood, *etc.* can be managed through internal or external models providing geospatial data that can be fed into the simulation.

The main features that the SWORD constructive simulation provides are:

- autonomous friendly, neutral and friendly units, who are able to understand and interpret operational orders given by a human operator and adapt to situation encountered during their mission, without any additional human intervention. Officers can command their subordinate units as they would on the battlefield instead of issuing waypoints and oversimplified orders to unrealistic units unaware of tactics, techniques and procedures:

  - wayfinding mode, taking into account tactical motivations and terrain features and existing infrastructure;
  - management of unit perception, each unit sharing its knowledge of a situation with its allies;
  - realistic behavior during movement, combat or support to other units, with consideration for human factors, like morale, fatigue and experience (specified);
  - non-determinist combat resolution, each unit evaluates local situation to determine if it wants to engage in combat, its optimal target, dependent on its Rules of Engagement (ROE), mission, weapons, force ratio, etc. (with the optional setting random factor).

- comprehensive management of the chain of command, allowing operators to control various echelons. For instance, company-level orders are automatically broken down into platoon-level orders for each platoon included in this company, to fulfill their missions using their respective capacities, ensuring maximal realism;
- management of military specific branches and a large brand of emergency services, civilian populations and irregular entities. These non-military units interact dynamically with military units regarding the course of action, allowing the scripting of complex scenarios through pre-recorded orders;
- mission timelines, boundaries, phase lines and objectives are directly taken into account by simulated units to establish a doctrine-compliant course of action;
- intuitive concepts and user-friendly interface can be fully translated to native language;
- comprehensive and customizable after action review, with many reporting and scoring tools;
- compatibility with various communication formats allowing interoperability with entity-level simulations and other constructive simulations;
- support of multiple terrain formats, raster data format and vector formats (like VMAP, DTED, USRP, GeoTIFF, ESRI shapefiles);

- integration of complex internal or external physical models, such as chemical, biological, radiological and nuclear (CBRN) cloud movement, propagation of fire, floods, and other natural disasters.

The SWORD and SWORD Analysis may use the same elaborative objects, entity, terrain, behavior, doctrine databases. They could be imported from other sources (file format typical) or defined with the use of set of build-in editors. Therefore, the user: simulation or modelling and simulation center, staff or decision-maker have all necessary tools to create, run, and analyze exercise, experiment or course of action in short time. To avoid accidental mismatch, especially when testing variants with different specifications of objects or entities, the analytical tool, and SWORD itself allow, allow creation separate databases to be defined and used. It is not necessary to build these new databases from scratch, because the software allows the user to simply make a copy of the existing one, and then modification of the new, while the original (reference) is kept unchanged.

The SWORD After Action Review (AAR) module, as in all constructive simulations, provides itself powerful validation and analysis tool. Available features in AAR are very similar to those available during the exercise. Since it is basically a replay, the units cannot be controlled in AAR. All events are recorded and the whole scenario can be replayed as a movie, step-by-step, forward or backward, with different user profiles, gauges, *etc*. Although, the course of events cannot be changed, the resulting video can be edited to meet the exercise director or the staff operator. This includes splitting the video (into different named and described sections) and removing those that are not required.

The software allows to run course of action comparison session, which looks like simultaneously running two different courses of action indicating the differences between them. What is interesting in this mode is that concurrent courses of action could be observed in real time (use of two screens is advisable for convenience). The presentation of the results in this mode is also designed accordingly to be able to visualize the comparison and differences between the two ways of mission execution (Fig. 1). This mode has the same after-action review capabilities with ability to present result of both CoAs at the same time.

Additionally, the SWORD Analysis have convenient export mode to other formats like pdf, MS Excel and PowerPoint, which helps staff to save time in preparation of the consolidated information package for briefings and decision sessions. What additionally helps in this task is that the SWORD software allows to make the stand-alone replay and ready-to-play package. It is a portable version of the exercise session, which can be copied on any mobile media (e.g. USB memory, DVD, memory card) and played on the computer without the SWORD software constructive simulation software installed, for example in the briefing rooms not connected to the external networks or in the restricted security-sensitive environments (after the proper check by the Information Security officers).

The presented case study software, as all similar systems, has limitations. It depends on the hardware capabilities (computing power of workstation, link bandwidth, networking equipment) and philosophy driving the development (usually

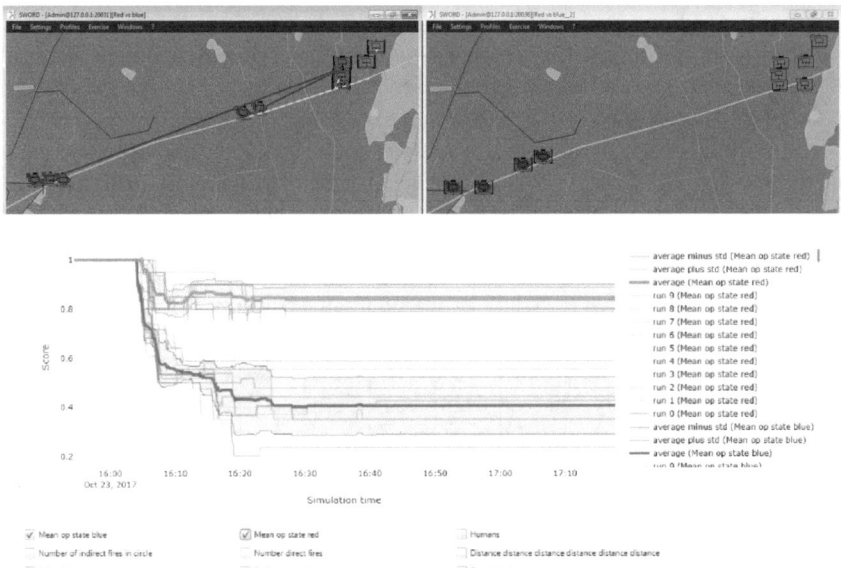

Fig. 1 Visual (top) and statistical (bottom) representation of the result of a course of action (variant) execution, Masa Group materials, 2020

inspired by the main user). Therefore, for a software producer, there is large space for improvement and development of this very promising technical solution, especially in an ability to interlink with various battle/information management systems and accepting live input data delivered by remote nodes and sensors, as well as possibilities to be fed by the data management systems (employing machine learning, neural networks, data farming, *etc.*). Additionally, the behavior algorithms and new simulation models (*e.g.* natural disasters, disease spread and effect, CBRN, space, jamming, electronic warfare, cyber) should be constantly developed to replicate the reality. The producers should also include ability to accept the feed from or federation with the social media and information operations (INFOOPS) simulators.

4 Discussion

The implementation of the artificial intelligence did not exclude the possibility to run the computer assisted exercise (CAX) in more traditional way with operators and workstation dedicated to the separate unit. Scalability and features of the simulation software, as well as the robustness of their databases, allow the simulation of various military and non-military scenarios that provide decision-makers support in their work.

Including AI-powered simulation software to battle management systems provides staffs and commanders with a support tool that can process input data and present the possible results in a split of a second, allowing them to see the possible end state and decide which course of action to choose.

It is also important to indicate that such systems must have access to sufficient, timely and detailed data. Aside from inputs provided by operators from the units, a BMS collects data in automated manner from connected nodes and sensors, such as unmanned aerial and ground vehicles (UAV/UGV) deployable sensors, satellites, aerial imagery, intelligence, surveillance and reconnaissance (ISR) assets, signal intelligence (SIGINT) assets, radars, or even wearables like Internet of Things (IoT). Such amount of information needs engines and protocols that automatically categorize, catalogues, index, and process it, and disseminate the result deciding what to send where. Additionally, the complexity of modern and future operations requires data from other sources that could supplement own databases and provide details on the events that were not predicted and detailed in the existing systems. The search engines need to know what the decision-maker or staff personnel needs to know about the object, event, system, without bothering the busy personnel with too many questions or requiring to fill elaborated forms. Therefore, such decision-making support systems should look into the smart protocols, systems and solutions, such as neural networks, machine learning or data farming to name only the few.

The neural network is a very simplified model of the human brain. It is a type of data processing, inspired by biological neurons, that converts between complex objects (such as audio and video) and tokens suitable for conventional data processing. It is made up of a large number of information processing elements called neurons. They are connected with each other in a certain way, which describes the type of the neural network: one-way single or multi-layered networks, recursive networks, cellular networks. They collect data and passes it on, use connections between neurons for searching, (which is used in learning process) and collect conclusions, *i.e.* analysis results.

A neural network can consist of any number of layers. In information technology, a neural network is hardware or software (or both) that is modelled after the operation of neurons in the human brain. Typically, a neural network is made up of many layers. The first layer, as in the case of images recorded, for example, by the optic nerves of a human—goes to raw input data. Each subsequent layer receives data resulting from data processing in the previous layer. What the last layer produces is the so-called system output. A neural network functions like the human brain: each neuron carries out its own simple calculations, and the network made up of all neurons multiplies the potential of these calculations. Neural networks used in artificial intelligence are organized on the same principle, but with one exception, to perform a specific task. The connections between neurons can be adjusted accordingly. Neural network technology has many practical applications. It is widely used for handwriting recognition, speech-to-text conversion, weather forecasting and facial recognition (Definition of Decision Intelligence; Definition of Neural Network).

Moreover, there is new term called decision intelligence, which is a practical realm that covers a wide range of decision-making techniques, combining many traditional and advanced disciplines to design, model, match, execute, monitor and adjust models and decision-making processes. These disciplines include decision management (including advanced nondeterministic techniques such as agent-based systems) and decision support, as well as descriptive, diagnostic and predictive techniques.

Data science on the other hand provides additional solutions, such as data mining, data farming, big data. They allow collection of data from various sources, so-called "white intelligence", open sources and classified databases, coping with multiple formats and categories of data, categorization of data. They might search and select the data from dedicated sources/databases, or all available open sources.

Specifically, data farming is a simulation-based methodology that supports decision-making throughout the development, analysis and refinement of courses of action. By performing many simulations runs, a huge variety of alternatives can be explored, analyzed and visualized to allow decision-makers improved situational awareness and to make more informed and robust decisions. It intends to allow decision-makers in the domains of defense planning, operations, training and capability development to reduce uncertainty resulting in more robust solutions (NATO, 2018).

Thanks to AI algorithms and cross-referencing/cross-checking the systems help with fake news and/or unconfirmed information distinguishing. It creates the challenge for categorization, prioritization, and computing. It is consuming both power and time. In dynamic situations commanders wishing to have the answers and analysis in a split second, which can be delivered by the data farming, AI, machine learning solutions.

There are of course doubts and threats that comes with the use of any technology, and especially with disruptive and not fully tested solutions. Tendency to overreliance on the technology can create situation where too busy commanders depend their decisions on the "cold, machine calculations", which although sophisticated in their design, lack social competences and so-called human dimension. Therefore, due to the time pressure, commanders/decision-makers tend to accept so-called "best" courses of action based on the raw data and calculations. It is hard for a machine to decide subtle humane factors that might dramatically change the course of decision.

Additionally, it is always a question of the security of the system and information flow. How to protect it against cyber-attacks and electronic warfare impact? How resilient they could be? Those questions would need to be addressed when discussing on the scale of implementation of the AI-powered decision-support systems.

Research findings reveal that, AI's support for the decision-making process in military operations is already a reality and will continue to develop. What we argue for in here, is that modern technologies are becoming necessary to support the activities of soldiers, commanders and staffs. However, the question is how far AI should develop. When it comes to the area of decision support, the authors agree that there is room for Artificial Intelligence, however, the decision is up to humans. Moving to another (not necessarily higher) stage of development will result in the decision not supporting the decision-making process but replacing it with AI algorithm. Today it is difficult to make educated statements that AI can threaten humanity because there are only futuristic images contained in novels. The unknown that we are facing applies to both biological and human thinking process, that will not achieve the quality and speed of future technologies. Computers, artificial intelligence, robots and autonomous systems will create an environment too complex and fast for humans to follow, much less direct. Gradually, probably unnoticed, automatic systems will run more and more efficiently, better than humans, so humans will become passive observers (Latiff, 2017). Looking further, we would place humans as the weakest link in the battle structure, taking away their responsibilities and possibilities of being independent in their thinking and taking actions.

Some of the decision-making authority military would like to delegate to the machines claiming that the procedures are understood and followed by the machines. An example could be the swarm of drones, that autonomously operates and decides what and when to attack or when particular "UAV member" of the swarm should go back to base. Is humankind willing to pass this choice to the machines equipped with the machine learning or neural networks capabilities? Probably not. Observing science fiction and the influence of technology on the development of civilization, it can be said that many of the futuristic visions have been realized and hves a potential to be realized.

The question of which leadership process should be automated by computers and which should be performed by human leaders leads to investigations not only of operations process but also human values and morals. The actions of AI devoid of human objectivity and humanism may change or even distort the dimension of future armed conflicts. It is worth to pose the question: Where in future armed conflicts, treated as counteracting aggression (taking fair and defense wars into account), is the place for comradeship, brotherhood, patriotism and humanism? According to Latiff (2017) by building a working machine that can come closer to human behavior and demonstrate elementary moral decision-making we might not know what to do with people being involved in the fight. This question opens a discussion on personal involvement into the conflict and social interaction between members. The key concept of camaraderie becomes meaningless concept (Latiff, 2017).

5   Conclusion

The use of artificial intelligence in combat operations has already become a reality. Currently, it supports the collection of information, its processing and suggesting the choice of appropriate action by soldiers or their commanders. The final decision belongs still to the human being who uses modern technologies, because we are dealing with artificial intelligence at its initial stage of development. There are many indications that in the future we will be dealing with independent AI units, making independent decisions, without direct human interference. This scenario raises many questions and concerns, especially of ethical nature.

Aspect is the principle of humanism and the understanding of dependencies in people's decision-making. Not every conflict situation has to end with the use of force. The experience of many armed conflicts shows that people behave differently, depending on the specific situation. History provides examples where well-equipped and superior military units made decisions to surrender even though they had the ability to fight (were combat effective). Is therefore the artificial intelligence in such situations a substitute for human behavior? Is it able to learn human sensitivity and empathy? Freedman (2017) highlights the importance of human involvement in the battles, with emphasis on ethical reasoning while making decisions, not only mathematic calculations. He argues that in the conflict scenario the objective was not to follow pre-design scenario, but to take action and respond to unfolded events. Following on that we might think again about it in leadership context. If computers take over leaders tasks, the right problem to focus on wouldn't be to fight against the technology advances, but rather how to automation can be designed in a way that generates work environments conducive to human satisfaction and well-being (Wescheand and Sonderegger, 2019; Visvizi & Lytras, 2019).

Whether it is cloud computing, machine learning, neural networks, AI-powered simulations, data farming or big data, the decision-making process can easily employ them. The complexity of multi-domain, multilateral, multi-agent operations conducted in the modern world requires enormous amount of information collected, processed and presented in short amount of time and in synthetic form so the decision makers can select the best course of action. There will always be a danger of making machines too independent or concern over the ethical nature of the "cold comput-ing". Moral dilemma over the choice of sending human soldier or unmanned vehicle to execute the task, where the human factor plays a crucial role, will persist as the pivotal issue. There is no question that we can cope with issues and dilemmas that commanders face during contemporary operations without support of the disruptive technologies supporting decision-making process.

The technology solutions for decision-making support are either available on the market and ready to use or in various stages of development ready to be delivered. One of the biggest challenges however, is the mentality of the decision-makers. The education system, especially in the realm of professional military education (PME), although discussing the possibility of use of modern technology solutions in support of decision-making, is not promoting in their actual implementation in daily staff and commander activity, both in time of peace and conflict. The standard military decision-making process cycle has a lot of space for those data science and simulation systems. Mainly in planning phase, but also in the following phases of the cycle, where the future orders can be evaluated, specifics honed, current operations tracked, the follow-on courses of action played and adjusted, the alternative/parallel plans developed, so when the situation changes there are possible solutions ready. Nevertheless the advantage the question arises: will robots replace humans in making decisions, and if so, will they be able to consider humanitarianism in their decision-making?

References

Agrawal, N. (2020). Modeling enablers of knowledge management process using multi criteria decision making approach. *VINE Journal of Information and Knowledge Management Systems.* https://doi.org/10.1108/VJIKMS-08-2019-0122

CfALL. (2015). MDMP Lessons and Best Practices, Handbook 15–06, March 2015, Fort leaven-worth : Center for army lessons learnt (CfALL). https://usacac.army.mil/organizations/mccoe/call/publication/15-06

Clemen, R. T. (2001). Naturalistic decision making and decision analysis. *Journal of Behavioral Decision Making, 14*(5), 359–361. https://doi.org/10.1002/bdm.385

Definition of Decision Intelligence—Gartner Information Technology Glossary. (n.d.). Gartner. https://www.gartner.com/en/information-technology/glossary/decision-intelligence

Definition of Neural Network—Gartner Information Technology Glossary. (n.d.). Gartner. https://www.gartner.com/en/information-technology/glossary/neural-net-or-neural-network

Freedman, L. (2017). In *The future of war: A history* (1st ed.). PublicAffairs.

Gonzales, C., Perny, P., & Dubus, J. P. (2011). Decision making with multiple objectives using GAI networks. *Artificial Intelligence, 175*(7–8), 1153–1179. https://doi.org/10.1016/j.artint.2010.11.020

Hamilton, R. F., & Herwig, H. H. (2004). Decisions for war, 1914–1917 (Abridged ed.). Cambridge University Press

Hodson, D. D. (2009). *Performance analysis of live-virtual-constructive and distributed virtual simulations: Defining requirements in terms of temporal consistency.* Air Force University of Technology.

Latiff, R. H. (2017). In *Future war: Preparing for the new global battlefield.* Knopf.

Machińska H., & Malinowski A. (1984). Wprowadzenie do technik decyzyjnych i organizatorskich [Introduction to decision-making and organizational techniques].

Mitchell, T. M. (1997). *Machine learning.* McGraw-Hill.

National Incident Management System, (2004). U.S. Department of Homeland Security. March 1.2004.

NATO Science and Technology Board statement on 18 September 2018. https://www.sto.nato.int/Pages/news.aspx

PREMIUM: Czechs put to the SWORD—Training—Shephard Media. (2020, October 23). Shephardmedia. https://www.shephardmedia.com/news/training-simulation/premium-czechs-put-sword/

Schraagen, J. M., & van de Ven, J. G. M. (2008). Improving decision making in crisis response through critical thinking support. *Journal of Cognitive Engineering and Decision Making, 2*(4), 311–327. https://doi.org/10.1518/155534308x377801

Sieci neuronowe. (n.d.). Sztuczna Inteligencja [Neural networks. Artificial Intelligence]. https://www.sztucznainteligencja.org.pl/definicja/sieci-neuronowe/

Simulation: Live, virtual and constructive. (n.d.). TNO. https://www.tno.nl/en/focus-areas/defence-safety-security/roadmaps/operations-human-factors/simulation-live-virtual-and-constructive/

Spector, L. (2006). Evolution of artificial intelligence. *Artificial Intelligence, 170*(18), 1251–1253. https://doi.org/10.1016/j.artint.2006.10.009

Ścibiorek, Z. (2003). *Podejmowanie decyzji [Decision making process].* PWN.

Ścibiorek, Z. (2018). *Uwarunkowania procesu decyzyjnego w niemilitarnych zdarzeniach nadzwyczajnych [Determinants of the decision-making process in non-military emergency situations].* PWN.

Toner, C. (2005). Moral issues in military decision making. *Journal of Military Ethics, 4*(2), 149–152. https://doi.org/10.1080/15027570510030879

Triantaphyllou, E., Shu, B., Nieto, S. S., & Ray, T. (1998). Multi-criteria decision making: An operations research approach. In J. G. Webster (Ed.), *Encyclopedia of electrical and electronics engineering* (Vol. 15, pp. 175–186). Wiley.

Visvizi, A., & Lytras, M. . (2019) 'Politics & ICT: Mechanisms, dynamics, implications. In Visvizi, A., & Lytras, M. D. (Eds.), *Politics and technology in the post-truth era.*

Vroom, V. (1973). Leadership and decision-making. *Organizational Dynamics, 28*(4), 82–94.

Wachowiak, P. (2001). Profesjonalny menedżer. Umiejętność pełnienia ról kierowniczych [Professional manager. Ability to perform managerial roles], Warszawa. Difin.

Wesche & Sonderegger. (2019). When computer take the lead: The automation of leadership. In *Computers in human behavior* (Vol 101, pp. 197–209). Elsevier. https://doi.org/10.1016/j.chb.2019.07.027

Wiśniewski, B., Kozioł, J., & Falecki, J. (2017). Podejmowanie decyzji w sytuacjach kryzysowych [Decision-making process in crisis situations], Szczytno. WSPol.

Wrzosek, M. (2018). Wojny przyszłości [The wars of the future], Warszawa. Fronda.

Zdyb, M. (1993). Istota decyzji [The essense of the decision], Lublin.

Wojciech Horyń Ph.D. in Security Science, Associated Professor, the Polish Armed Forces Colonel (ret,). Professor of the Faculty of Security Sciences at the Military University of Land Forces. Former Dean in the University of Business in Wrocław. Member of the board of the Academic Society for Andragogy, member of European Association for Security and member of the Polish Society of Safety Sciences. Research areas: security policy, crisis management, education of security, andragogy, education. Email: wojciech.horyn@awl.edu.pl ORCID: 0000-0002-9887-5889.

Marcin Bielewicz Ph.D. in History, Lieutenant Colonel in the Polish Armed Forces, Military Assistant to the Commander, NATO Joint Forces Training Centre, Associate Editor of the International Journal of Security Studies of the University of North Georgia, United States. Graduate of the US Army Infantry School and the US Naval Postgraduate School. Member of the NATO Modelling and Simulation Group. Participant of research and academic international internship programs. Research areas: international security affairs, armed conflicts, modelling and simulation, disruptive technologies, transatlantic relations, military education and training. Email: bielewiczmarcin@yahoo.com ORCID: 0000-0002-7941-1747.

Adam Joks Ph.D. in Security Science, Major General in the Polish Armed Forces, Commander of the NATO Joint Force Training Centre. Member of the Stakeholders Board of the Faculty of Political Science and International Studies at the University of Warsaw. Graduate of the George C. Marshall European Center for Security Studies, the Strategic Studies at the US Army War College, and the Generals, Flag Officers and Ambassadors' Course at the NATO Defense College. Research areas: international and national security, crisis response operations and conflict management, warfare development and innovation, strategic leadership. Email: joksadam@gmail.com ORCID: 0000-0002-2410-6040.

AI, Decision Making, and the Public Sector

Artificial Intelligence Systems in the Decision-Making Process

Marian Kopczewski⊙

Abstract The theory and practice of management has long been looking for methods and tools enabling the implementation of management functions in an organization based on the decision-making process based on modern information technologies. ICT has been successfully supporting management processes in organizations for many years, including in the form of integrated management information systems (eg MRP or ERP classes), decision support systems (SWD) or systems known as Business Intelligence for some time. The essence and purpose of the chapter is therefore to identify the impact of AI on the decision-making process based on the theory of fuzzy sets and fuzzy logic explaining the ways in which decision systems operate.

Keywords Artificial intelligence · Decision process · Business intelligesnce · Theory of fuzzy sets

1 Introduction

Since the last decades of the last century there have been processes described as an information revolution, the meaning of which some scientists compare to the invention and introduction of writing. The *information revolution* is characterized by the emergence of large and complex organizational structures (companies, agencies, institutions), advanced information processing technologies used by these structures, and the interpenetration of human and machine activities in production processes. Every revolution has positive and negative social effects. As in the *industrial revolution* of the nineteenth century, when the mass introduction of machines increased the production capacity of various kinds of goods and at the same time caused a significant increase in unemployment, in the information revolution the elimination of the need for people to do primitive and burdensome work (they are replaced by machines and computers) is observed (Rutkowska, 1999; Thomas et al., 2017; Visvizi & Lytras,

M. Kopczewski (✉)
Military University of Land Forces, Wrocław, Poland
e-mail: m.bodziany@interia.pl

2019). It may also lead to structural unemployment and new types of employment (Malik et al., 2021). From the social point of view, the negative effects are similar in both revolutions, however, for the economy, having highly processed information, obtaining new information in a short period of time, its skillful use in the processes of quick decision making and management determines to a large extent the success of undertakings and further development of institutions.

The article aims to present practical examples that gathering, and processing information is not an end in itself but a sub-goal (tool) in the decision-making process facilitating and accelerating effective management of processes, human resources, enterprises, institutions etc. The main tool supporting decision-making is a computer system of information processing, included in the decision-making system, where the final link is a human being (head of department, director, president, etc.). The type of decision making in an enterprise, similarly to the organizational structure of an enterprise or institution, has a hierarchical character with a pyramid structure, as shown in Fig. 1.

At the lower level of this pyramid, basic operational information such as technological process documentation, accounting of financial operations, balance sheets in the company's departments, project documentation, etc. are collected and processed, characterized by large amount and detail. After the analysis and appropriate processing of this information, its essential parts are transferred in a synthetic form to a higher organizational level and are placed (data) in the so-called *Information Management System* (ISZ) (Yager & Filev, 1995). The ISZ contains databases created based on modern technologies and file management systems with appropriate languages to facilitate access to data and files. At this organizational level, information is no longer as numerous and detailed as at the basic level, but its quality and value are much greater. ISZ assists the heads of company or institution departments

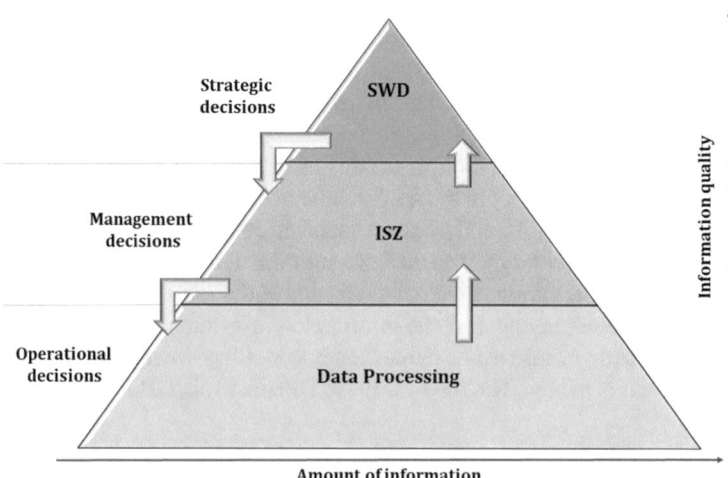

Fig. 1 The interdependence of decision types (Kopczewski, 2018)

in making decisions at this level. After the analysis of the information contained in ISZ, the most important part of it is transferred in a synthetic form to the *Decision Support System* (SWD) (Levithan & Yossi, 2018), which assists the top management in making strategic decisions. SWDs are organized and created using artificial intelligence methods, in which an important role is played by appropriate representation of knowledge, gathering information in knowledge bases, inclusion of automatic inference based on the possessed knowledge (cf. Visvizi & Lytras, 2021). In this process the uncertainty about the fragments of knowledge and uncertainty in the inference needs to be taken into account. Similarly, *heuristics* has to be considered. This is necessary to account for the rules of 'common sense' and strategies reducing the area of problem space searches and accelerating solutions to the problems posed. Like in expert systems, an essential issue in SWD is to equip them with an appropriate interface that should be user-friendly, as high-level directors and managers are generally not computer scientists or proficient in advanced technologies of obtaining and processing information technologies.

2 Decision-Making

Decision-making is one of the links in the decision making process, which includes the following stages: *defining the problem and specifying the goal* (recognizing and defining the essence of the decision situation), *examining the variants of the choice of the decision and predicting the consequences* (defining and identifying alternative options), *choosing the optimal variant* (the best one according to specific criteria), *sensitivity analysis* (examining the impact of changes in the optimal variant when conditions change) and *implementing the decision* (Białko, 2000; Tefekci 2018). Decision making itself is the act of choosing one 'best', i.e., the most effective option from among many alternatives (Shahsavarani & Abadi, 2015; Bonatti et al., 2009). The effectiveness should be understood as not only increasing company's profits, sales, etc., but also minimizing possible losses or the most beneficial withdrawal from business activity. While predicting the consequences of deciding, models are used, i.e., simplified descriptions of processes, phenomena, etc., and they can be of deterministic or probabilistic type. In general, decisions can be divided into two categories: programmed and not programmed.

Programmed decisions have a clear structure or are repeated regularly. For example: the sales manager knows from experience that he/she must keep a specific weekly stock of a product; hence, he/she can introduce an automatic refill system if the stock is less than a specific one. Similarly, a lecturer giving lectures on some Saturdays at a branch will know what to do if he/she 'receives a call' from the branch manager on the preceding days. This kind of decision making can be fully programmed.

Decisions that are not programmed appear less frequently and not regularly and have no clear structure. Examples of such decisions can be opening a new point of sale, merging institutions, discontinuing business activity, setting a price for a new product, etc. In cases where such decisions are made, *heuristics* related to the decision makers' intuition, 'common sense', and experience play an important role (Graham & Jones, 1988; Thomas et al., 2017).

The conditions under which decisions are made are characterized by states of certainty, risk, and uncertainty. A state of certainty is a situation in which the deciding person knows with great certainty the available choices and their conditions; e.g., buying a mid-range brand passenger car: Ford, Opel, or Citroen. Choosing one of the options does not carry the risk that the choice was incorrect. The state of risk is a situation in which the decision-maker knows the benefits and costs of the available options only with some estimated certainty (probability); e.g., buying one of the abovementioned new models whose operating conditions are unknown or a used car. Decision-making in conditions of uncertainty is a situation in which the person making the decision does not know all the choices, the risks associated with each of them, or the possible consequences arising from them. In order to make the most appropriate decisions in uncertainty, decision-makers should obtain as much useful information as possible about the future decision, be able to analyze it quickly, 'filter' information that is of little use, and get some suggestions for a more favorable choice of one of the possible options. IT decision-making support systems play an important role in such cases (Brian, 2018; Lytras & Visvizi, 2021).

An important feature of the decision-making process is the uncertainty of the rationale on which it is based and the uncertainty of the final decision as the result of the inference. Therefore, one of the most critical issues in the operation of the decision support system is effective processing of uncertain and imprecise information. For these purposes, calculation methods based on probability calculus and others such as *certainty factors* (Cox, 1994), fuzzy sets and fuzzy logic, non-monotonic logic, etc. are applied. Among them, a method based on fuzzy systems, where uncertainty is encoded in the fuzzy statements and fuzzy rules and processed based on generalized (fuzzy) laws of logic (e.g., *Modus Ponendo Ponens*) during the execution of individual rules, is very convenient (Białko, 2000; PoAI, 2018; Sarirete et al., 2021).

3 Representation and Processing of Imprecise Knowledge by Fuzzy Systems

In fuzzy systems, knowledge is represented by fuzzy statements (fuzzy facts) and fuzzy rules, also called fuzzy unconditional and conditional rules, respectively. On the other hand, rules contain *linguistic* variables whose values are expressed in verbal (linguistic) form and are represented as fuzzy sets.

The fuzzy sets differ from the classic sets in that their values do not change violently ('sharp') when passing from one to the next, but permeate, and the belonging of a given element of the x variable to the set A is determined by the value of the belonging function $\mu_A(x)$. An example of graphical representation of three values

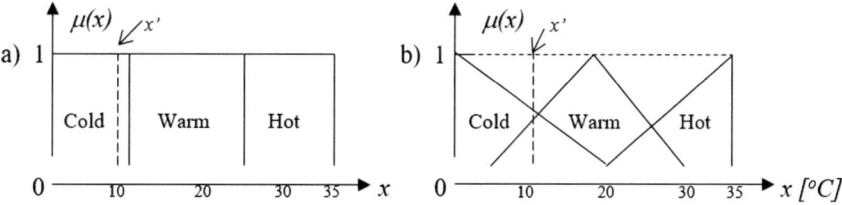

Fig. 2 Values (thermae) of the x = temperature variable: **a** classic, **b** fuzzy

(thermae) of the linguistic variable *temperature*, in classic and fuzzy form, is shown in Fig. 2 (Białko, 2000; PoAI, 2018).

E.g., the specific temperature $x' = 10$ °C is, in the classic case, definitely cold, because $\mu_{cold}(10) = 1$ and $\mu_{warm}(10) = 0$, while in the fuzzy case—more *cold* and less *warm*, because $\mu_{cold}(10) \cong 0.5$ and $\mu_{warm}(10) \cong 0.4$. As one can see, the values of the affiliation function can be treated as the factors of certainty of a state; the higher the value $\mu(x)$ the lower the uncertainty.

The shapes of the fuzzy sets and their ranges are defined by the designer of the fuzzy system and can be presented graphically (as in Fig. 2), analytically—in the form of the functional description $\mu(x) = f(x)$ defining the shape of the set, and in the form of a set of discrete pairs of values $\mu(x_i)/x_i$; e.g. the fuzzy set $\mu_{cold}(x)$ from Fig. 2b can be presented as (Białko, 2000; PoAI, 2018):

$$\mu_{cold}(x) = \{1/0, 0.9/2, 0.8/4, 0.7/6, 0.6/8, 0.5/$$
$$10, 0.4/12, 0.3/14, 0.2/16, 0.1/18, 0/20\}$$

where a diagonal line does not mean dividing but linking numbers in pairs. Typical shapes of fuzzy sets are: triangle, trapezoid, bell curve, e.g., Gauss, and so called singleton, i.e., a set having the value $\mu_{A'}(x') = 1$ only for x', and zero for other x.

The shapes of the sets can be changed accordingly using the so-called modifiers. The most commonly used ones include the concentration operator associated with the concepts: *very* or *extraordinary*, which is defined as:

$$con(\mu_A(x)) = (\mu_A(x))^2$$

and the dilution operator associated with concepts *slightly* or *a little*, which is defined as:

$$dil(\mu_A(x)) = (\mu_A(x))^{1/2}$$

For the $\mu_{cold}(x)$ set defined above, the concentrated set $\mu_{verycold}(x)$ will be:

$$con(\mu_{B-Z}(x)) = \{1/0, 0.81/2, 0.64/4, 0.49/6, 0.36/8,$$

Fig. 3 Approximate fuzzy
set diagrams: cold, very cold,
slightly cold

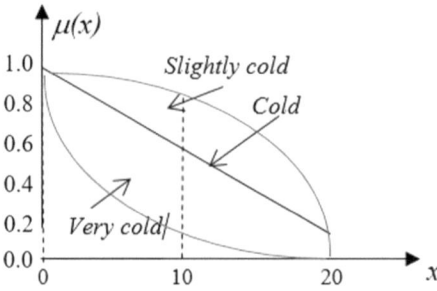

$$0.25/10, 0.16/12, 0.09/14, 0.04/16, 0.01/18, 0/20\}$$

and the diluted set $\mu_{slightly\ cold}(x)$ will be:

$$dil(\mu_{N-Z}(x)) = \{1/0, 0.95/2, 0.89/4, 0.84/6, 0.77/8,$$
$$0.71/10, 0.63/12, 0.54/14, 0.45/16, 0.32/18, 0/20\}$$

Diagrams of such defined fuzzy sets: *Cold, Very cold, Slightly cold* are shown in Fig. 3 (Białko, 2000).

The basic operations on fuzzy sets include:

- *intersection of sets,* marked with ' ∩ ', corresponding in the logic to the conjunction '∧' or the logical conjunction AND,
- *combination of sets,* denoted by ' ∪ ', corresponding in logic to the alternative '∨' or the logical conjunction OR,
- *completion of the set,* marked with ' ~ ', corresponding in logic to the negation '¬' or the operator NOT.

The standard intersection of the two fuzzy sets $A \cap B$, represented by the $\mu_A(x)$ and $\mu_B(x)$ belonging functions, is defined as:

$$\mu_{A \cap B}(x) = MIN(\mu_A(x), \mu_B(x)) = MIN(A, B),$$

and the standard connection $A \cup B$, – as:

$$\mu_{A \cup B}(x) = MAX(\mu_A(x), \mu_B(x)) = MAX(A, B)$$

The fuzzy completion to the set A is defined as

$$\sim \mu_A(x) = 1 - \mu_A(x)$$

The graphical representation of the standard intersection and connection and completion of the fuzzy sets is shown in Fig. 4 (Białko, 2000; PoAI, 2018).

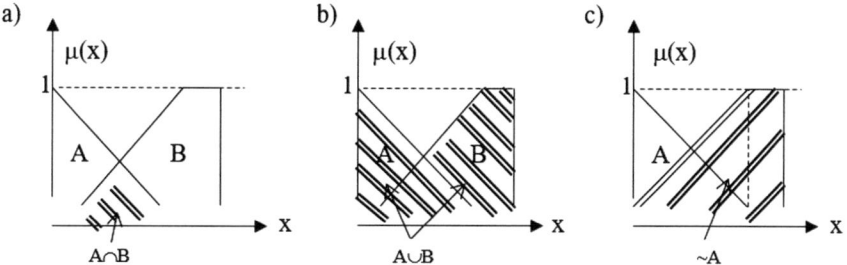

Fig. 4 The graphical representation (standard): **a** intersection, **b** connection, **c** completion

In general, the intersection of fuzzy sets A, B, is defined using a function called *a triangular norm (T-norm)*:

$$\mu_{A \cap B}(x) = T(\mu_A(\mu), \mu_B(\mu)) = T(A, B)$$

while the combination of sets A, B—using the function called the *triangular T-norm or S-norm*:

$$\mu_{A \cup B}(x) = S(\mu_A(x), \mu_B(x)) = S(A, B)$$

Examples of more frequently used triangular standards are shown in Table 1 (Białko, 2000).

Due to the values obtained after using the individual *T-norm* and *S-norm*, they can be ranked as follows:

$$R\text{-}ogr \leq Il\text{-}E \leq Il\text{-}alg \leq Il\text{-}H \leq Min \leq$$
$$Max \leq S\text{-}H \leq S\text{-}alg \leq S\text{-}E \leq S\text{-}ogr$$

For example, for $A = 0.4$ i $B = 0.8$ and $A = B = 0.5$ the results shown will be obtained in Table 2.

Table 1 Examples of T-norm and S-norm

$T(A,B)$(intersection)	$S(A,B)$(connection)
MIN(A,B) (minimum, Min)	MAX(A,B) (maximum, Max)
AB/(A+B−AB) (Hamacher product, Il-H)	(A+B−2AB)/(1−AB) (Hamacher sum, S–H)
AB (Algebraic product, Il-alg)	A+B−AB (Algebraic sum, S-alg)
AB/{1 + (1-A)(1-B)} (Einstein product,, Il-E)	(A+B)/(1+AB) (Einstein sum, S-E)
MAX(0, A+B−1) (Limited difference, R-ogr)	MIN(1, A+B) (Limited sum, S-ogr)

Table 2 Triangular norm values for different A and B

Norm	R-ogr	Il-E	Il-alg	Il-H	Min	Max	S-H	S-alg	S-E	S-ogr
$A = 0.4, B = 0.8$	0.2	0.286	0.32	0.36	0.4	0.8	0.823	0.88	0.91	1.0
$A = B = 0.5$	0.0	0.2	0.25	0.333	0.5	0.5	0.667	0.75	0.8	1.0

In addition to triangular standards, operators with intermediate properties between *MIN* and *MAX* are also suitable. These are the so-called *average operators*, which include, among others, the following.

$$Harmonic\ mean : 2AB/(A + B)$$
$$Geometric\ mean : (AB)^{1/2}$$
$$Arithmetic\ mean : (A + B)/2$$

and so called *parameterized gamma* operators, which are used to define operators: *fuzzy-MIN/MAX* and *fuzzy-Product/Algebraic sum*:

$$Fuzzy\text{-}MIN/MAX : \gamma MIN(A, B) + (1 - \gamma)MAX(A, B),\ \gamma \in [0, 1]$$
$$Fuzzy\text{-}Product/Algebraic\ sum : \gamma AB + \{(1 - \gamma)(A + B - AB),\ \gamma \in [0, 1]$$

For the value of the parameter $\gamma = 1$, the *fuzzy-MIN/MAX* corresponds to the operator *MIN* and the *fuzzy-Product/Algebraic sum* corresponds to the operator *algebraic product*; for $\gamma = 0$, the first gamma operator corresponds to the operator *MAX* and the second to the operator *algebraic sum*, while for $\gamma = 0.5$, both operators correspond to the *arithmetic mean*.

The order of averaging operators over triangular norms is as follows:

$$T\text{-}norms \leq Averaging\ operators \leq S\text{-}norms$$

or more specifically:

$$MIN \leq Harm.mean \leq Geom.mean \leq Arithm.mean \leq MAX$$
$$MIN \leq Fuzzy\text{-}MIN/MAX \leq MAX$$
$$Il\text{-}alg \leq Fuzzy\text{-}Il/Alg.sum. \leq S\text{-}alg$$

Although the most commonly used triangular standards in fuzzy systems are *MIN, MAX* and *Algebraic Product* (Mentel, 1995), other standards and averaging operators are also used in fuzzy decision support systems when these standards lead to ambiguous results or raise doubts about the accuracy of the suggested decision.

The presented intersection and connection operations concern fuzzy sets defined on the same base set X, i.e. they concern thermae of the same linguistic variable x. On the other hand, if we want to perform these operations between the thermae of different linguistic variables defined on different base *X, Y, Z,* etc. sets, then we use fuzzy relationships.

Like the fuzzy set described by values $\mu(x_i)$ spanning over the *x*-axis with points x_i, i.e., by pairs $\{\mu(x_i)/x_i\}$, the relation between sets is described by pairs $\mu(x_i),\mu(y_j)$ spanning over the Cartesian plane x_i,y_j created by the product of the Cartesian vectors with elements x_i,y_j.

Fuzzy relationships are used to model the implications of fuzzy rules and the relationships of statements (facts) occurring independently as unconditional rules or as premise for complex conditional rules. By executing them, unconditional and conditional rules are used to simulate inferences and decisions (Kopczewski et al., 2012; Gidley, 2017).

Examples of statements (facts) or unconditional rules can be:

Speed is Low, Price is High, etc., also saved as:
Speed = Low, Price = High.

Simple conditional rules have the form:

If (x = A) To (y = B), or:
If (premise), then (conclusion), e.g.:
If (price is Low), then (sale is Large) or:
If (price = Low), then (sale = High)

while complex conditional rules consist of a number of premises connected by a logical consistency *AND*:

$$If\,(premise-1)(premise-2)\ldots(premise-N)then(conclusion)$$

which means that all the conditions should be met in order to reach a conclusion; an example is given here:

$$If\,(price\text{-}home = Acceptable)(height\text{-}tax = Acceptable)$$
$$then(purchase\ decision = big)$$

Inference in fuzzy systems is based on the generalized law of *Modus Ponendo Ponens* logic, which is written in form:

$$\{[(x\ is\ A) \Rightarrow (y\ is\ B)] \wedge \left(x\ is\ A^{'}\right)\} \Rightarrow \left(y\ is\ B^{'}\right)$$

which reads as:

If for a given fuzzy implication (a conditional rule), a fuzzy fact (set) A' associated with its premise A is entered, then a modified fuzzy set B' associated with its conclusion B is obtained.

For example, let's consider the rule:

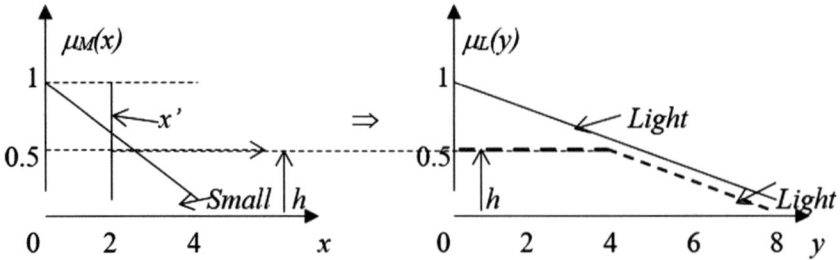

Fig. 5 Exemplary inference for the rule: If x = M then y = L, for x' = 2 (singleton)

$$If (the\ size\text{-}object\ is\ Small) then (the\ weight\text{-}object\ is\ Light)$$

where: *size-object* ≡ *x, weight-object* ≡ *y*.

When adopting the following sets of shapes: *Small(x), Light(s)*, as shown in Fig. 5, the relationship between them in the implication (rule execution) is $\mu_{R,M\Rightarrow L}(x,y)$ using *T-norm* (e.g. operator *MIN*):

$$\mu_{R,M\Rightarrow L}(x, y) = MIN(\mu_M(x), \mu_L(y))$$

The fuzzy inference can be easily explained graphically, as shown in Fig. 5 (Białko, 2000): however, for calculation purposes, computer programs use matrix account and the relationship concept.

Assuming the "sharp" value of the singleton $\mu_{sing}(x')$ at the position $x = x' = 2$ as the input quantity, one calculates its intersection with the set of the input variable *Small x*, obtaining the value *h* called the *rule ignition* factor:

$$h = MIN(\mu_{sing}(x'), \mu_{small}(x))$$

The process of converting a "sharp" value *x'* into a value *h* is called *blurring* or *fuzzification*. Then, as the result of the *implication*, the value *h* is transferred to the diagram of the thermae *Light* of the variable *y*, creating a modified set *Light'(y)* marked with a bold line in Fig. 5. The modified set of *Light'(y)* is obtained by calculating the intersection:

$$mu_{Light'}(y) = MIN(h, mu_{Light}(y))$$

To get a specific "sharp" value *y'* of the output variable, the *Light'(y)* set should be converted to a single value. This is done through a process of *sharpening* or *defuzzification*. One of the most common methods of *defuzzification* is Center of Gravity method performed according to the integral formula:

$$y' = \frac{\int y\mu_{Light'}(y)dy}{\int \mu_{Light'}(y)dy}$$

Since fuzzy systems do not require much precision in calculations, a simplification is used to discredit values y and replace integrals with finite sums:

$$y' = \frac{\sum_0^n y_i \, h_i}{\sum_0^n h_i}$$

where h_i is the set height for value y_i.

Among other ways of defuzzification, the *height* belong to the simplest and is often used. It consists in taking the value y' corresponding to the place on the axis y where the value of the modified output set is maximum. It is used when there is a clear maximum in the output set or when the output variables of the fuzzy system are singletones.

4 Fuzzy Decision Support Systems

Although fuzzy systems are mainly used in automation devices, they are also applied in other branches of science and economy, such as decision support, automation of information extraction from databases, object classification, pattern recognition, etc. (Kopczewski et al., 2012).

An example of decision support can be a simple fuzzy system that advises in buying a house outside the city center, with several, sometimes opposing, criteria. These criteria may be: acceptable—not excessive or small price, acceptable—not high value of the annual property tax, acceptable—small distance from the workplace, acceptable—small distance from school (transporting children), possible attractiveness of the house (comfort of finishing, pleasant environment, etc.) (Kopczewski et al., 2012).

The listed criteria may be formulated as the so-called unconditional statements (facts), also called unconditional rules, and the form of:

R1: The price should be low (acceptable)
R2: The tax should be small (acceptable)
R3: Distance from work should be small (acceptable)
R4: Distance from school should be small (acceptable)
R5: Attractiveness should be high (acceptable).

The further execution corresponds to the use of *T-norm*, e.g., as an operator MIN. The above unconditional rules can be presented in the form of fuzzy sets, determined by a real estate specialist, as in Fig. 6 (Białko, 2000).

Let us assume that there are four houses 'A', 'B', 'C' and 'D', with values of criteria whose values are presented in Table 3 (these values are also shown in Fig. 6).

Whereas in Table 4, there are values of the membership function $\mu(x_i)$ corresponding to the degree of acceptability of the given x_i criteria for individual houses (determined based on the fuzzy sets from Fig. 6), together with the resulting degree

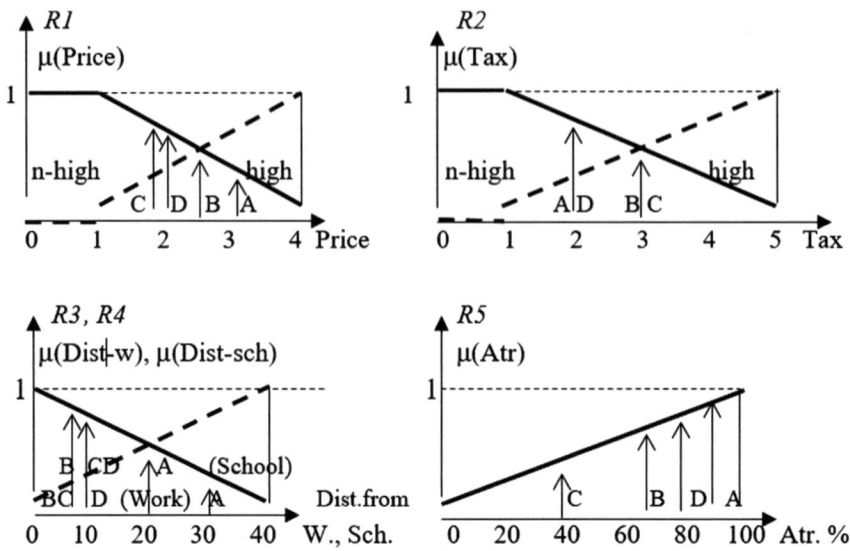

Fig. 6 Fuzzy sets corresponding to the unconditional rules R1-R5

Table 3 Criteria values for individual houses

House	Price × 1000 PLN	Tax × 1000 PLN	Dist-work km	Dist-school km	Atractiveness %
A	300	2	30	20	90
B	250	3	8	8	70
C	175	3	8	10	40
D	200	2	10	10	80

Table 4 Degrees of acceptability of criteria and resultant degrees of houses 'A' – 'D'

House	μ(Price)	μ(Tax)	μ(Dist-w)	μ(Dist -sch)	μ(Atrac)	$\mu_w = MIN(\mu_i)$
A	0.33	0.75	0.25	0.50	0.90	0.25
B	0.50	0.50	0.80	0.80	0.70	0.50
C	0.75	0.50	0.80	0.75	0.40	0.40
D	0.667	0.75	0.75	0.75	0.80	0.667

of acceptability $\mu_w = MIN(\mu(x_i))$ being the minimum values (operation *MIN*) of the degree of acceptability of the criteria for the given house.

With such criteria, the house recommended for purchase (among the available: A, B, C, D) is the house with the highest resultant degree of acceptability $\mu_{MAXw} = MAX(\mu_{wi})$, i.e., house 'D'. In this example, the choice is obvious, because for this house it is clearly the largest. However, it may happen that these coefficients are

Table 5 Degrees of acceptability of houses 'E' – 'H'

House	μ(Price)	μ(Tax)	μ(Dist-w)	μ(Dist -sch)	μ(Atrac)	$\mu_w = MIN(\mu_i)$
E	0.4	0.7	0.4	0.8	0.5	0.4
F	0.7	0.4	0.4	0.7	0.6	0.4
G	0.6	0.3	0.7	0.9	0.8	0.3
H	0.8	0.8	0.6	0.35	0.7	0.35

the same or comparable; so what selection criteria should be adopted then? Let us consider another example with houses E, F, G, and H with the degree of acceptability of the criteria as in Table 5, where the values of the resultant coefficients μ_{wi}, when accepting the operator MIN, are comparable and two of them are equal.

It should be emphasized that when accepting the operator MIN in the execution of rules, only the smallest values of the coefficients are considered, while omitting others. The use of other *T-norm*, e.g., the *product-algebraic* operator or averaging operator, e.g., *geometric mean*, which for many arguments takes the form:

$$\mu(x_1, \ x_2, \ ... \ x_n) = \prod_1^n (x_i)^{1/n}$$

will allow for taking into of the values of other coefficients μ_i, as shown in Table 6.

From Table 6 one can see that taking other coefficients into account indicates that the decision to buy a house 'F' is slightly more advantageous than 'E'. In addition, it is important that the acceptability levels of houses 'G' and 'H' are clearly higher than those of houses 'E' and 'F', which is due to higher values of other coefficients not previously taken into account. This effect is called *compensation* (Klir et al., 1977).

In a similar way, decision-making when buying or selling other objects or articles can be can supported.

Another example of decision-making support is the fuzzy system for determining the price of a new product introduced to the market. Determining the price of a new product is done taking into account many imprecise and uncertain factors such as demand for the product, prices of similar products from other manufacturers (competitors), costs of direct production, transport and storage costs, costs of advertising, the desire to obtain the highest possible profits, etc. In this example, only

Table 6 The resulting acceptability levels for operators: MIN, Il-Alg., Geom.mean

House	$\mu_w = MIN(\mu_i)$	$\mu_w = Il\text{-}Alg(\mu_i)$	$\mu_w = Alg.mean(\mu_i)$
E	0.4	0.0448	0.538
F	0.4	0.0470	0.543
G	0.3	0.0907	0.618
H	0.35	0.0941	0.623

a few factors have been taken into account for simplicity purposes: the desire to make large profits (high price) while ensuring a high level of sales (low price), the desire to have the sales price of the product equal to approximately twice the cost of its manufacture and to be similar to the price of another manufacturer (competitor) (Samuelson, Marks 1988).

The pricing strategy is based on the recommendations of the managers of the relevant plant departments, which may be contradictory. For example, the finance manager suggests that the price of the product should be as high as possible for high profits, while the sales and marketing department suggests that the price should be low to make sales mass and control the market. Based on these suggestions, two unconditional rules (statements) can be created:

R1: The price should be high R2: The price should be low

The production manager, on the other hand, suggests that the sales price should be approximately twice the direct production cost of the product; this can be expressed by the third unconditional rule (Samuelson, Marks 1988):

R3: The price should be about 2 manufacturing costs

In addition, the marketing manager also suggests that if the price of the competitor is not high, the product price should be approximately equal to the competitor's price; hence, the fourth rule is conditional and takes the form:

R4: If the price of the competition is not high, then
the price should be approximately equal to the competition price

It should be emphasized that the consecutive unconditional rules (statements or facts) are treated as multiple conditions occurring in the left parts of the conditional rules (premises) and, thus, are connected by logical conjunctions AND. Therefore, operations *MIN* are used for their subsequent execution, i.e., intersections of the corresponding fuzzy sets are determined.

After creating the rules, the fuzzy sets are to be defined: *high price, low price* in the assumed price range, e.g., *10 ÷ 30 price units*, and the input data accepted: *cost-producing* and *price-competition*, and the fuzzy sets defined: *about 2 × cost-producing* and *approximately equal to the price-competition*. Let us assume the following data:

Cost-producing: 9 units
Price-competition: 24 units.

Fuzzy sets corresponding to unconditional rules *R1, R2, R3,* determined by an expert, can be as shown in Fig. 7 (Białko, 2000).

If only the rules *R1* and *R2* existed, their execution would give an intersection of these sets with the shape of a shaded isosceles triangle; the defuzzification of this shape would give a price equal to 20 units. Further implementation of the rule *R3: about 2 production costs*, treated as a fuzzy number $2 \times 9 = 18$ with a triangular set, would result in the set intersection, with the previous resultant set (shaded) with the shape hatched with dotted lines. To obtain the resultant shape of the product

Fig. 7 The fuzzy sets corresponding to the rules: R1—high price, R2—low price, R3—about 2 × production costs

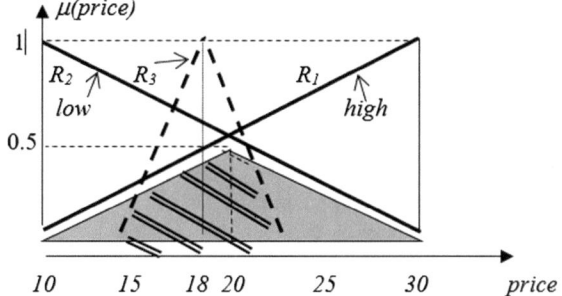

price, after the execution of the conditional rules *R4*, it is necessary to first make a conclusion and determine the shape of the set of the rule conclusions, and then to make an aggregation, i.e., a combination (operation *MAX*) of this set with the resultant set of the execution of the rules *R1, R2, R3* (the hatched set in Fig. 7).

The rule *R4* execution, assuming an *approximate price-competition* = 24 of a 'triangular' fuzzy number, is shown in Fig. 8.

The shape of the resulting fuzzy set, after combining the shaded set from Fig. 8 with the hatched set from Fig. 7, is shown in Fig. 9.

After creating the resulting fuzzy set, one can perform the defuzzification to get the resulting price of the product. If the defuzzification was done with the *maximum*

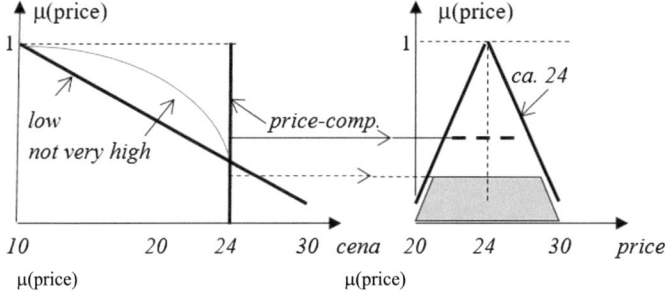

Fig. 8 Illustration of R4 rule execution (price- competition = 24)

Fig. 9 The resulting fuzzy set after completing R1 ÷ R4 rules

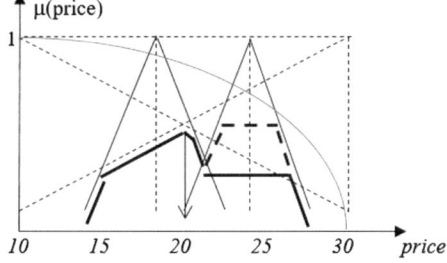

method, the resultant price would be 20 (the value of the price corresponding to the highest point of the resultant set from Fig. 9—marked with an arrow). In this case, the difference between the resultant prices is small. However, with other input price values and other definitions of fuzzy numbers and rules, these values may differ significantly. For example, if one modifies the premise of an R4 rule to a form: *If the price-competition is not high, then...* its execution will be slightly different (the shapes marked with dashed lines in Figs. 8 and 9) and the price will be higher (Samuelson & Marks, 1988).

Another important element of supporting the decision-making process is the assessment of the risk associated with the project to be implemented (research, investments, etc.), as failure to achieve the project objectives exposes the institution to serious losses. Such assessment is based on many factors such as project duration, financing, complexity, etc. An example of fuzzy rules related to the assessment may be as follows (in fact, there may be several or several dozen rules) (Kuo, 2001).

R1: If the project duration is long this risk is increased
R2: If the project team is large this risk is increased
R3: If the project funding is small this risk is increased

Conclusion on risk assessment is different from the typical application of fuzzy systems. The so-called *scaled monotonic inference* is used, resulting in a "sharp" risk value rather than a fuzzy set that is subject to defuzzification. In scaled monotonic inference, the used fuzzy sets can only be monotonously increasing or monotonously decreasing, and the conclusions of all rules are identical.

An example of scaled monotonic inference is shown in Fig. 10.

Assuming the input data of the rule rationale as e.g. (Yager & Filev, 1955).

project duration = 3 months, project team = 10 people, %-project funding = 40,

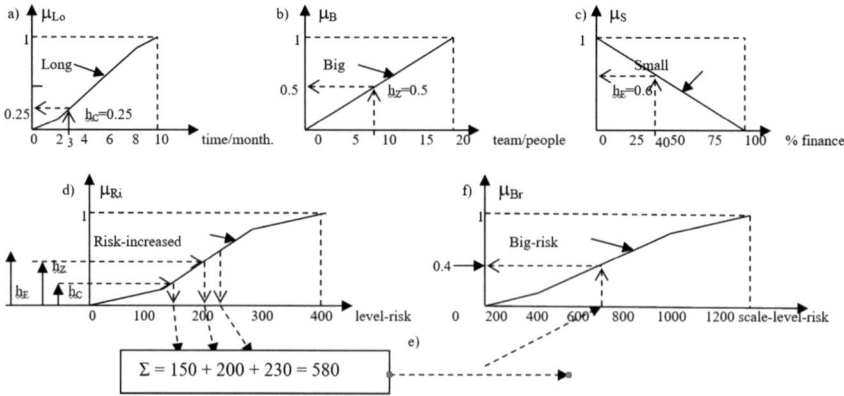

Fig. 10 Principle of monotonic conclusion: **a, b, c** sets of fuzzy premises of rules R1, R2, R3, **d** fuzzy conclusion of rules, **e** summation of conclusions, **f** scaled conclusion

The values of the coefficients of the belonging function (marked by h_i in Fig. 10a, b, c, d) are obtained: $h_C = 0.25$, $h_Z = 0.5$, $h_F = 0.6$. Then, according to the principle of inference, these values are transferred to the *augmented-risk* fuzzy set (see Fig. 10d) (Zimmermann, 1994), which is the conclusion of all considered rules. Here, instead of defuzzification, in accordance with the principle of monotonic inference, the intersections of h_i values and corresponding fuzzy sets are projected on the degree-risk axis, obtaining respectively values (the scale is arbitrary): 150, 200, 230 representing the degree of risk resulting from individual rules. Next, the risk degree is cumulated by summing up these numbers and obtaining the value $\Sigma = 580$ (Fig. 10e). Next, the risk level is scaled - assuming that the highest scaled value should correspond to the highest value of the risk level x the number of applied rules (here, with the value of 400 (Fig. 10d) and three rules the value of 1200 (Fig. 10f) is obtained). Next, for cumulative value $\Sigma = 580$ the value of belonging function $\mu_{Br} = 0.4$ (Fig. 10f) is obtained. Finally, this value is multiplied by the highest value of the rule conclusion range (here $0.4 \times 400 = 160$ (Fig. 10d)) to 'match' it to the level of risk degrees resulting from individual rules. This is the end of the risk assessment.

The presented examples show the convenience of using fuzzy set theory and fuzzy logic in IT systems to support decision making. This is beneficial when processing large amounts of inaccurate and uncertain information.

5   Conclusions

The theory and practice of management has long been looking for methods and tools enabling the implementation of management functions in an organization based on the decision making process. An important feature of the decision-making process is the uncertainty surrounding the premises on which it is based, as well as the uncertainty of the final decision made as a result of inference. Therefore, one of the most crucial issues in the operation of a decision support system is the effective processing of uncertain and imprecise information. For these purposes, computational methods based on the calculus of probability and other such as: Certainty factors, fuzzy sets and fuzzy logic, non-monotonic logic, etc. are used. fuzzy, in which the uncertainty is encoded in fuzzy statements and fuzzy rules and is conveniently processed based on the generalized (fuzzy) laws of logic.

An inseparable part of the decision support system is a module based on elements of artificial intelligence. Thanks to it, it is possible to analyze data and draw conclusions in a way close to the way people think—using uncertain or fuzzy data, analogies and learning methods. Decision support systems are increasingly used in many areas of life, especially in solving problems for which there are no algorithmic solutions. These systems are modern IT tools that provide managers with reliable information, both at the strategic and operational levels. System users receive instruments that quickly help to transform data dispersed in various IT systems of the company into coherent information.

According to the definition, an intelligent decision support system is characterized by the ability to learn and adapt to the needs of participants in the decision-making process. The techniques used in artificial intelligence systems include: expert systems with a knowledge base and intelligent systems (machine learning—the method of induction of decision trees). Today, many companies do not have integrated intelligent decision support systems. Generally using incomplete monitoring systems, they collect data and perform simple analyzes of that data. Currently, neither prognostic models nor decision support systems controlling the information supply network are used in the existing organizations. The introduction of an intelligent decision support system for control allows for correct profit prediction, which is presented in the article on selected examples.

Acknowledgements The article was written as part of the development project of the Ministry of Science and Higher Education "LESS" (Logistics Engineering of Security and Safety), 2017–2018, Lodz University of Technology - team member, and a research project financed by the Ministry of National Defense "Support information technology in making decisions in the event of threats ", 2017—2018, AWL Wrocław—project manager.

References

Białko, M. (2000). *Basic properties of neural networks and hybrid expert systems*. University of Technology.

Bootle, R. (2019). *Work, wealth and prosperity in the age of robots,* Hodder & Stoughton.

Bonatti, E., Kuchukhidze, G., & Zamarian, L. (2009). Decision making in ambiguous and risky situations after unilateral temporal lobe epilepsy surgery. *Epilepsy & Behavior, 14,* 665–673.

Brian, P. (2018). *Artificial intelligence and ethics: twelve areas of interest*. University of the Pacific, Stockton.

Cox, E. (1994). *The fuzzy systems handbook*. Academic Press.

Graham, I., & Jones, P. (1988). *Knowledge, Uncertainty and Decision*. Chapman a. Hall.

Gidley, J. M. (2017). *Future. A very short introduction*. Oxford University Press.

Klir, G., Clair, U., & Yuan, B. (1977). *Fuzzy Set Theory*. Prentice Hall, Upper Side River.

Leviathan, Y., & Yossi, M. (2018). *Google Duplex: An AI System for Accomplishing Real-World Tasks Over the Phone.* https:// ai.googleblog.com/2018/05/duplex-ai-system-for-natural-conversation.html.

Lytras, M. D., Visvizi, A. (2021). Artificial intelligence and cognitive computing: methods, technologies, systems, applications and policy making. *Sustainability, 13,* 3598https://doi.org/10.3390/su13073598

Mendel, J., (1995), *Fuzzy Logic Systems for Engineering*, Proc. IEEE.

Kopczewski, M., Tobolski, M., & Dudzik, I., (2012). *Models of information management in an organization*. In: Public management. In W. Kieżun, J. Wołejszo, & S. Sirko (Eds.), Warsaw: AON Publishing House.

Kopczewski, M. (2018). *Tools for Smart Intelligenc in the process of decision (review work - analysis of solutions)*. In *SIA Omniscviptum Publlishing—online monograph*.

Kuo, R. J. (2001). A sales forecasting system based on fuzzy neural network with initial weights generated by genetic algorithm. *European Journal of Operational Research,129*(3).

Lingras, P. (1998). Applications of rough patterns, rough sets in knowledge discovery 2. In L. Polkowski, & A. Skowron, (Eds.), New York: Physica-Verlag, Heidelberg.

Malik, R., Visvizi, A., & Skrzek-Lubasińska, M. (2021). (2021) The gig economy: current issues, the debate, and the new avenues of research. *Sustainability, 13*, 5023. https://doi.org/10.3390/su1 3095023

Mendel, J., (1995). *Fuzzy Logic Systems for Engineering*, Proc. IEEE.

PoAI (2018). Safety-Critical Working Group Meeting, Partnership on AI (PoAI): San Francisco, USA, https:// www.partnershiponai.org/thematic -pillars/. (21.01.2021)

Rutkowska, D., Piliński, M., & Rutkowski, L. (1999). *Neural networks, genetic algorithms and fuzzy systems*. PWN.

Sarirete, A., Balfagih, Z., Brahimi, T., Lytras, M.D., Visvizi, A. (2021). Artificial intelligence and machine learning research: towards digital transformation at a global scale. J Ambient Intell Human Comput (2021). https://doi.org/10.1007/s12652-021-03168-y

Shahsavarani, A. M., & Abadi, E. A. M. (2015). The bases, principles, and methods of decision-making: a review of literature, *International Journal of Medical Reviews, 2*(1), 214–225.

Tufekci, Z., (2018). *YouTube, the Great Radicalizer*, The New York Times. https://www.nytimes.com/2018/03/10/opinion/sunday/youtube-politics-rad-ical.html.

Thomas, A., Kasenberg, D., & Scheutz, M. (2017) *Value Alignment or Misalignment—What Will Keep Systems Accountable?*, Association for the Advancement of Artificial Intelligence. https:// hrilab.tufts.edu/publications/aaai17-alignment.pdf.

Samuelson, W., & Marks, S. (1998). *Managerial economics*. Polskie Wydawnictwo Ekonomiczne.

Visvizi, A., & Lytras, M. D. (2019). Politics & ICT: mechanisms, dynamics, implications. In A. Visvizi, & M. D. Lytras (Eds.), *Politics and Technology in the Post-Truth Era, Bingley*. Emerald Publishing.

Visvizi, A., & Lytras, M. D. (Eds.). (2021). *Artificial intelligence and cognitive computing: methods*. MDPI Publishing.

Yager, R., & Filev, D. (1955). *Fundamentals of modeling and fuzzy control*, WNT.

Zimmermann, H. (1994), *Fuzzy Set Theory*, Kluver Academic Publ.

Marian Kopczewski PhD, professor, researcher at the Academy of Land Forces in Wroclaw, Poland, Faculty of Security Sciences. His expertise covers security, crisis management, as well as the use of information systems in management and teaching, as well as national and internal security systems, including European and Euro-Atlantic political and military integration processes. He is the author and co-author of several publications related to national security and IT systems. Participant and active speaker at many conferences and scientific meetings, Prof. Kopczewski manages scientific and research works on a domestic and foreign scale. He is a member of the Polish Society for Production Management, the Polish Society for Safety Sciences and the Polish Association of Creative Teachers.

Artificial Intelligence and the Public Sector: The Case of Accounting

Gennaro Maione⬤ and **Giulia Leoni**⬤

Abstract The implementation of AI-enhanced systems in the business context offers many benefits. However, in accounting studies, AI is still configured as an almost unexplored frontier. This gap is even more clear for public sector accounting, given the relatively small number of scientific contributions dedicated to the topic. In light of these considerations, the work aims to highlight the limiting factors and the enabling drivers for AI the public sector accounting. To this end, a qualitative approach was applied to study the answers provided by a sample of 45 managers, placed at the head of the accounting office of some Italian municipalities. The questions were prepared in the form of semi-structured interviews, developed by enucleating and adapting the key concepts related to the five attributes that, according to the Innovation Diffusion Theory, characterize every innovation process: relative advantage; compatibility; complexity; trialability; and observability. The findings suggest that Italian public sector accounting is experiencing a transition, placing itself halfway between the "early adopter" and the "early majority", that is, in a phase in which current technologies begin to be perceived as outdated and not worthy of further investment, whilst the process of spreading new AI-based technologies appears interesting but still immature.

Keywords Artificial intelligence (AI) · Public sector accounting ·
Innovation diffusion theory · Qualitative research · Semi-structured interview

G. Maione (✉)
University of Salerno, Fisciano, Italy
e-mail: gmaione@unisa.it

G. Leoni
Polytechnic University of Marche, Ancona, Italy
e-mail: g.leoni@pm.univpm.it

© The Author(s), under exclusive license to Springer Nature Switzerland AG 2021 131
A. Visvizi and M. Bodziany (eds.), *Artificial Intelligence and Its Contexts*, Advanced
Sciences and Technologies for Security Applications,
https://doi.org/10.1007/978-3-030-88972-2_9

1 Introduction

In the business environment, the implementation of technology provides many advantages (Agostino & Arnaboldi, 2016; Bracci & Vagnoni, 2006). In particular, among all "new" technologies, Artificial Intelligence (AI)-based systems offer several benefits (Visvizi and Lytras, 2019a, 2019b, 2018), such as the ability to automate the performance of repetitive tasks, upload documents automatically, classify items, increase the accuracy of the information generated by processing the collected data, encourage spontaneous machine learning, suggest the path to obtain the best possible result. The perception of these significant advantages is increasingly evident, so much so that the spread of AI is a constantly growing phenomenon, at the level of both operating practice and scientific literature (Arasteh et al., 2016; D'aniello et al., 2016; Kokina & Davenport, 2017; Troisi et al., 2020, 2021). In fact, in recent years, many scholars from various disciplinary sectors have been orienting their research interests towards the investigation of the effects deriving from the use of AI in companies (Ciasullo et al., 2018; De Maio et al., 2015; Polese et al., 2017; Ukpong et al., 2019). However, in accounting studies, especially in the Italian ones, AI is configured as an almost unexplored frontier, although at the level of operational practice the first applications—less advanced than those of today—date back to the mid-1980s (Baldwin et al., 2006). This gap in the literature is even more clear concerning the accounting of public bodies, given the relatively small number of scientific contributions dedicated to the topic (Wirtz et al., 2019). Nonetheless, in the face of this gnoseological delay, there is empirical evidence that underlines how the use of AI brings significant benefits for accountants (Salawu & Moloi, 2020; Ukpong et al., 2019). These researches are conducted mostly in the private sector. On the other hand, some researches highlight the advantages deriving from the adoption of intelligent systems in the public sector (Sun and Medal, 2019; de Sousa et al., 2019; Visvizi et al., 2018a, 2018b; Chui et al., 2018) but the number of accounting studies is negligible. In light of these considerations, to fill the gap found in the literature, the work aims to highlight the limiting factors and the enabling drivers for AI the public sector accounting. To this aim, this study follows a qualitative survey approach, based on the analysis of the answers provided by a sample of 45 managers, placed at the head of the accounting office of some Italian municipalities. The questions were prepared in the form of semi-structured interviews, developed by enucleating and adapting the key concepts related to the five attributes that, according to the theory originally proposed by Everett Rogers (2010), characterize every process of spreading an innovation: relative advantage; compatibility; complexity; trialability; and observability. The argument in this chapter is structured as follows. The following section offers an overview of the debate on AI, especially as regards its role in the domain of business accounting and the public sector. Then, the research design is described with reference to the methodology used for the construction and administration of the interviews, the sampling procedure, and data collection; Subsequently, the results emerged are discussed. Finally, theoretical-managerial implications and conclusive considerations about the ideas for future research are debated.

2 Artificial Intelligence for Business Accounting

AI-based technologies are employed in both public and private sector. The progressive adoption of AI systems represents a fundamental step for the future of the accounting (Elliott, 1992): who deals with accounting needs new technological tools to increase the cost-effectiveness of the activities carried out, especially in terms of lower economic, temporal, and cognitive resources, as well as greater reliability of the decisions to be made (Issa et al., 2016). Yet, most researchers who, up to now, have deepened the topic of AI in support of accounting, do not have a wealth of knowledge and experience about intelligent IT systems, having rather a general background referable to information systems (Baldwin et al., 2006). Conversely, the possibility that expert AI researchers in the strict sense explore the implications in the accounting field appears unusual (Metaxiotis & Psarras, 2003). Hence, with some exceptions, in literature, there is a disconnection between the scientific domain of accounting and the technological domain of AI, although there is empirical evidence that underlines the profitable opportunities deriving from the interdisciplinarity of the two areas. Mosteanu and Face (2020), for example, through a survey aimed at identifying the effects deriving from the use of technology to support the drafting of accounting documents, claim that AI-based systems, combining financial information with IT skills, accelerate the digital transformation of accounting, reducing the probability of error determined by human intervention. In this regard, Sutton et al. (2016) state that recent technological evolution has made techniques based on AI algorithms capable of supplying considerable support to who makes assessments and decisions during and after the audit process. Similarly, Zemánková (2019) discusses the benefits arising from the widespread use of AI in accounting and, more specifically, in auditing, placing emphasis on the role played by the latest technologies in ensuring greater efficiency and integrity of audit revisions. Consistently, Mol and Yigitbasioglu (2019) recommend the use of AI as an approach to the automation of certain accounting procedures to increase the effectiveness of the business decision-making process. Similarly, Cheng and Roy (2011) implement a promising alternative approach (Evolutionary Fuzzy Support Vector Machine Inference Model for Time Series Data) consisting of a hybrid AI system focused on managing time-series data characteristics, which merges fuzzy logic, weighted support vector machines, and an advanced genetic algorithm to minimize the margin of error due to human inputs in the corporate cash flow forecasting processes. However, not all studies show positive results about the feeling of the effects deriving from the implementation of AI systems to support business accounting. In this regard, thanks to an exploratory research conducted in Australia, Kend and Nguyen (2020) highlight that, whilst, on the one hand, the approval towards experimenting of the synergies between intelligent systems and accounting is growing, on the other, there are still hesitations, relating, for instance, to the need for a common regulation for the various countries of the world.

2.1 Artificial Intelligence in the Public Sector

In the last decade, public sector companies appear increasingly interested in the use of AI as a viaticum for the effective and efficient exploitation of the resources and managerial skills necessary to implement sustainable policies in contexts characterized by a high degree of uncertainty (Mikhaylov et al., 2018). Whether private companies already use some intelligent tools and technologies in a widespread way, the public sector, despite a few years of delay, is starting a series of processes placing AI at the service of citizens, in contests like health, social, legal, tax, administrative, security, education, etc. (Boyd & Wilson, 2017; Desouza et al., 2020). In this direction, Wirtz et al. (2019) outline a conceptual approach to offer a broad overview of the use of AI-based instrumentation in the public sector, identifying ten possible application areas and four dimensions, attributable to the challenges that public companies will face shortly: security; specialized competence; data integration; and financial feasibility. Likewise, Desouza et al. (2020) share reflections and insights deriving from their experience in AI projects applied to the public sector, organizing the knowledge gained in 4 distinct but related thematic domains: data processing; implementation of technologies; organization of processes; adaptation to the environment. Similarly, de Sousa et al. (2019) carry out a review of the literature relating to scientific contributions dedicated to AI in the public sector, identifying the three main functions most frequently involved: general public service; economic affairs; and environmental protection. In addition to the studies focusing on the systematization of literature dedicated to AI in the public sphere, there are also empirical researches that assess some intelligent applications in the field. Androutsopoulou et al. (2019), for instance, in collaboration with three Greek government agencies—the Ministry of Finance, a social security organization, and a large local government organization—present a pragmatic approach aimed at the advanced exploitation of a specific AI technology, the chatbots, in the public sector, to contribute to the improvement of communication between government and citizens. The approach proposed by the authors is based on natural language processing, machine learning, and data mining technologies and exploits pre-existing data in various forms (such as documents containing laws and directives, data structured by the operating systems of government agencies, social media data, etc.). Similarly, Mbecke (2014) addresses the issue of improving communication in the South African public sector, using an artificial intelligence tool (Bayesian networks) that facilitates the definition, quantification, and combination of different factors that contribute to the provision of public services. By simulating different scenarios, the author promotes a theoretical approach as a methodological tool to allow public companies to adequately exploit AI-based communication technologies to improve the perception of the quality of services provided to citizens. Similarly, Sánchez (2019) provides evidence of how a new Blockchain-based AI paradigm for database management (Distributed Ledger Technologies) can be successfully implemented in public procurement governance, ensuring the achievement of a high level of transparency, integrity, autonomy, and speed. Consistently, Sun and Medal (2019) analyze a case of adoption of an artificial intelligence system (IBM Watson) in Chinese public health to map how three groups of stakeholders -government policymakers, hospital managers, and IT managers—perceive the challenges of AI adoption in the public sector, by defining four guidelines to facilitate the governance of public companies in the process of implementing AI systems: avoiding adopting shortsighted policies concerning the benefits of AI; implementing adaptive strategies to reconcile divergent views on AI; prioritizing the development of AI-based technologies and, more generally, data integration; focusing on AI governance rather than on AI generated by AI.

3 Research Design

3.1 Approach

Considering the peculiarity of the services offered, the characteristics of the stakeholders involved, and the social impact of the aims pursued, AI has interesting potential development, especially in the public sector. In this regard, to identify both the limiting factors and the drivers of AI for accounting in the public sector, this work is carried out by following a qualitative survey, based on the collection, analysis, and interpretation of unstructured and non-numerical data (Hennink et al., 2020). Given its flexibility, this approach is widely employed in the social sciences and, more particularly, by accounting scholars since it allows observing and treating even complex phenomena (Silverman, 2020), investigating gnoseological paradigms ranging from positivism to post-positivism, from the critical theories to constructivism (Guba & Lincoln, 2005). Through qualitative research, the researcher aims to understand not only what the unit of analysis (e.g. individual, private company, public body, etc.) thinks, believes or guesses but also the motivations underlying the relative opinion (Qu and Dumay, 2011). The qualitative approach is followed through the administration of semi-structured interviews. The choice to use this data collection technique (instead of open or structured interviews) is justified by the consideration according to which, although presenting a fixed track, the further development of the interview varies according to the answers progressively provided (Horton et al., 2004). In fact, by administering semi-structured interviews, the researcher can deepen some topics that spontaneously emerge and that could be useful for understanding the phenomenon investigated (Cohen et al., 2002).

3.2 Interview Construction

The basic scheme of the interviews administered has been defined by enucleating and adapting the key concepts related to the five attributes that, according to the theory originally proposed by Everett Rogers (2010), characterize every process of spreading an innovation: relative advantage; compatibility; complexity; trialability; and observability. According to the scholar, the diffusion, that is the process by which an innovation is transmitted over time to the members of a given social system, depends on the relative advantage, consisting in the perception that an innovation is considered to be better than the idea, program or product it intends to replace. To make this perception as positive, the compatibility requirement has to be also respected, which expresses the consistency of the proposed innovation concerning the values, experiences, and needs of potential users. The third attribute identified by Rogers, the complexity, acts in the opposite direction, expressing how difficult it is to understand and/or use an innovation. To this end, diffusion is favored by trialability, which indicates how much an innovation can be tested before making a commitment to adopt it in the future. Finally, according to the author, to materialize the dissemination process, an innovation must comply with the observability requirement, which reflects the degree to which an innovation provides tangible results that can be evaluated. The five key elements of the innovation diffusion process interact with each other and therefore they need to be considered as a whole (Rogers, 2002). It could happen, for example, that an innovation appears to be extremely complex, reducing its probability of being adopted and widespread, but, on the other hand, it could be very advantageous compared to current alternatives and compatible concerning the background or expectations of potential users (Van de Ven & Rogers, 1988). Based on this consideration, to identify the current constraints and possible future levers of the diffusion process of AI for accounting in the public sector, a semi-structured interview has been developed, consisting of five open-ended questions, one for each attribute, as indicated below: (1) "What are the benefits that come or could derive for accounting in the public sector from the use of AI-based tools?" (2) "Do you think that the public sector is ready to implement AI functionalities? (3) "Do you believe that those who deal with accounting in the public sector own or can acquire the skills necessary to understand and adequately use AI devices?" (4) "Do you think that AI can be experimented in the public sector in support of accounting?" (5) "In what terms the results obtained or obtainable through the diffusion of AI for accounting in the public sector should be evaluated?".

4   Analysis and Findings

Once the interview was structured, the next step was to collect data. To this end, the sampling procedure began with the sending of an email presenting the research project—context, objective, research questions -, and a request for membership to the heads of the administrative-accounting offices of all the municipalities of the province of Salerno. A total of 275 emails were sent because, although the municipalities identified were 158, the administrative and accounting offices of some of them have multiple email addresses. The managers of the offices of 74 municipalities responded to the first email, most of whom asked to receive the interview before expressing their willingness to join. A second email was sent containing a file with the interview to be administered and 57 municipalities agreed to take part in the project by responding to the interview. However, only 45 of them returned the updated file with their responses. Overall, the data collection phase lasted about 4 months, from 13 January to 11 May 2020. Subsequently, the collected data were analyzed by the two authors, who, at first, to avoid a possible mutual influence, acted separately, interpreting the answers based on their knowledge background concerning the Innovation Diffusion Theory (Rogers, 2010). Subsequently, the comparison between the authors became necessary, as well as appropriate, to better target the conceptualization of the limiting factors and drivers enabling the diffusion of AI for accounting in the public sector. However, the comparison revealed uniformity in the interpretation of the information extrapolated from the interviews and this aspect highlight the coherence of the research design concerning the link between the objective pursued and the results obtained. Appendix 1 includes the most significant extracts of the interviews administered, discussed in the following section.

5 Discussion

The work fits into the panorama of scientific contributions that attempt to foster the progressive acquisition of a more mature awareness about the limits and perspectives of new AI-based technologies, which could shortly revolutionize the role of accountants employed in the public sector. In this sense, the study provides several insights, potentially capable of generating implications for both researchers and professionals in the public sector. As for the theoretical implications, the work contributes to the enrichment of the state of the art focused on the diffusion of innovation in public sector accounting. In fact, the analysis provides empirical evidence of how Rogers' (2010) Innovation Diffusion Theory can be concretely interpreted in the public sector, allowing identifying the limiting factors (RQ1) and the enabling drivers (RQ2) of AI for accounting. With regard to limiting factors, the results highlight: an overall concern about the length of the period of adaptation to new technologies; the perceived risk of the existence of flaws in the IT systems capable of compromising the professionalism of the accountants; the lack of cultural schemes to face a challenge of this magnitude in an integrated way; and the scarce training offer on AI for accounting for public employees. Instead, about the identification of drivers enabling the spread of AI in support of public sector accounting, the analysis revealed: greater efficiency of services; facilitation of communication and interaction between public bodies and citizens; orientation of accountants' work towards more rewarding tasks; less effort for public employees in extracting significant information; and more effective and timely decisions. With reference to the managerial implications, the work allows becoming aware of the evolutionary stage of the diffusion process of AI in support of accounting in the public sector and, consequently, of the interventions to be programmed shortly. According to the results emerged, referring to Rogers' Innovation Diffusion Theory (2010), it is possible to suppose that public sector accounting is halfway between the "early adopter" and the "early majority" (see Fig. 1). This is not an unexpected result since, by its nature, the public sector is

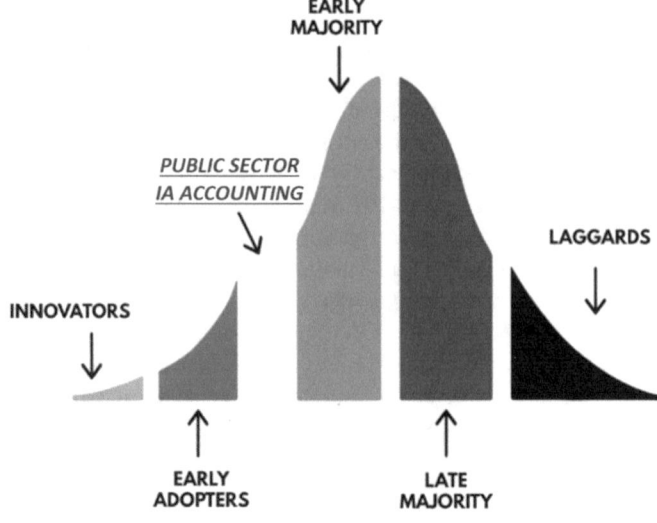

Fig. 1 The phases of the AI diffusion process for accounting in the public sector. *Source* Authors' elaboration

hardly configured as an "innovator" (Potts, 2009), especially because, usually, private companies initiate experiments capable of promoting the diffusion of innovation, subsequently and possibly also implemented in public institutions (Borins, 2001). Development strategies in the public sector hardly show a high-risk appetite, which is a typical feature of any innovative process: they are implemented only after the benefits deriving from innovation are tangible. This approach should not be understood as a lack of interest in innovation but, rather, as the intention to pay attention so that resources, often limited, are appropriately invested, avoiding waste and vain attempts of technological advancement.

6 Conclusions

Overall, the answers provided by the interviewees suggest that accounting in the Italian public sector is experiencing a transition period, that is, a phase in which current technologies begin to be perceived as outdated and not worthy of further investments, whilst the process of spreading new AI-based technologies appears interesting but still immature. Therefore, to stimulate the process of adapting accounting to the new perspectives that AI is progressively outlining, priority must be given to the development of guidelines for AI management (Sun & Medal, 2019), reconsidering in a coordinated way all the aspects that the diffusion of an innovation entails, in terms of updating legal regulations, culture, training, social policy, and so forth. Like any transition period, also in this circumstance, there are fears and uncertainties, increased by the fact that, in this case, the transition from traditional technology to AI for public sector accounting implies the almost complete replacement of an "intelligent" machine to man and this could be destabilizing. In conclusion, in light of the results emerged from the analysis and in consideration of the fact that, so far, still little has been done at the central and local government level to encourage the emergence of synergies between AI and accounting, the authors invite accounting scholars to provide their contribution, directing their research interests towards AI, being a largely unexplored frontier, able, whether properly studied, to offer significant advantages in terms of both scientific advancement and progress at the operating practice level. In this sense, this article is only a further step towards the acquisition of a more mature awareness of the benefits for accounting in the public sector deriving from the diffusion of AI-based technologies. Even because the work is not devoid of limits, mostly related to the sampling method. In fact, although there was no desire to use filters for the selection of accountants to be interviewed, for reasons related to the feasibility of the research, it was chosen to limit the selection to the heads of administrative-accounting offices of a single Italian province. This choice could affect the generalizability of the results. Therefore, in future research on the topic, the analysis will be carried out by extending the sample to verify the existence of possible analogies and/or differences between the resulting findings.

Appendix 1

The answers of the interviewees

Question	Extracts from the answers
(1) "What are the benefits that come or could derive for accounting in the public sector from the use of AI-based tools?"	• "AI will be the added value for every public company … as it will stimulate the efficiency of services and the expansion of skills … favoring the diffusion of useful innovation … for the benefit of both professionals and citizens, families, and "enterprises" • "AI applied to accounting will ensure a better experience for users … facilitating communication and interaction with public bodies" • "AI will open up a range of opportunities that will help guide accountants' work towards more rewarding tasks" • "Thanks to AI, those who deal with accounting in the public sector will be able to work more intelligently … managing to extract more significant information from moles of data, including unstructured data … facilitating the making of more effective and timely decisions"
(2) "Do you think that the public sector is ready to implement AI functionalities? "	• "Although numerous AI tools are already available, the public sector is at a primordial stage in a very long path of adjustment … which will require a huge investment in the near future." • "It is currently unimaginable to rely on AI to deal with the aspects characterizing the activities related to public sector accounting" • "It still takes a long time for the AI consolidation process in support of accounting to be considered completed… especially because the current information systems, implemented for several years now, still have flaws that risk compromising the professionalism of the accountants"

(continued)

(continued)

Question	Extracts from the answers
(3) "Do you believe that those who deal with accounting in the public sector own or can acquire the skills necessary to understand and adequately use AI devices?"	• "Public employees who deal with accounting do not have the cultural frameworks to face a challenge of this magnitude in an integrated manner yet" • "There is no real training in the Public Administration… and in accounting there is an even more pronounced skill gap" • "To date, there is no effective focus on the development of skills specifically dedicated to artificial intelligence… and this is leading to excessive delays in the transition from the pilot to the implementation phase" • "Accounting in the public sector is currently in full swing for the acquisition and exploitation of the experience necessary to generate added value linked to the use of AI-based technologies" • "At the beginning, every process of innovation frightens but, over the years, public sector accountants have shown that they know how to adapt to frequent organizational changes… and once again they will be able to assert their skills" • "The skills required to safely handle AI-based technologies cannot be acquired without difficulty but… with a view to continuous improvement, the accounting offices will be able to manage the process of the digital revolution that is investing the public sector" • "The mistake is believing that to use AI systems appropriately, those involved in accounting of the public sector need particularly technical IT skills… but it is not so because these aspects will be addressed by specialists in the digital sector. It will be sufficient to train the accountants in such a way as to allow them to effectively and efficiently govern the logic underlying the AI systems"
(4) "Do you think that AI can be experimented in the public sector in support of accounting?"	• "It is realistic to imagine that in a few years, at the most, AI will be a certainty for all public sector employees who deal with accounting" • "The world is evolving more and more rapidly and it is not possible to ignore the need to ensure the adaptation to new smart technologies… especially in crucial areas of the public sector, such as accounting… where efficiency will continue to represent one of the main objectives to pursue with determination" • "The spread in the Public Administration of AI-based devices for accounting should not be discussed but, rather, promoted… because whether the lack of investment in the development and implementation of new intelligent technologies risks generating serious repercussions on future generations" • "AI, accounting, and the public sector are a complicated but necessary trinomial… because it is unthinkable to imagine a future orphan of new technologies, increasingly adherent to the expectations of those who use and benefit from them"

(continued)

(continued)

Question	Extracts from the answers
(5) "In what terms the results obtained or obtainable through the diffusion of AI for accounting in the public sector should be evaluated?"	• "The actual ability to automate the routine must be taken into consideration, to be assessed in terms of both effectiveness and efficiency" • "Accounting in the public sector will benefit from the contribution offered by AI only whether it is possible to guarantee the achievement of a high degree of accuracy, precision, and timeliness of the information generated" • "The empathy of new AI-based technologies must be considered because, especially in the public sector, the objective is to encourage the improvement of the citizens' life quality" • "One aspect that should not be underestimated is the assessment of the cost–benefit ratio: investing in AI to support accounting can prove to be a particularly expensive activity… with benefits that are difficult to find immediately" • "The impact on employment must be taken into account because, if on the one hand, the AI allows performing tasks automatically and faster than the operations performed manually, on the other hand, this feature could jeopardize a large number of jobs in public sector accounting "

References

Agostino, D., & Arnaboldi, M. (2016). A measurement framework for assessing the contribution of social media to public engagement: An empirical analysis on Facebook. *Public Management Review, 18*(9), 1289–1307.

Androutsopoulou, A., Karacapilidis, N., Loukis, E., & Charalabidis, Y. (2019). Transforming the communication between citizens and government through AI-guided chatbots. *Government Information Quarterly, 36*(2), 358–367.

Arasteh, H., Hosseinnezhad, V., Loia, V., Tommasetti, A., Troisi, O., Shafie-khah, M. & Siano, P., (2016, June.). Iot-based smart cities: a survey. In *2016 IEEE 16th International Conference on Environment and Electrical Engineering (EEEIC)* (pp. 1–6). IEEE.

Baldwin, A. A., Brown, C. E., & Trinkle, B. S. (2006). Opportunities for artificial intelligence development in the accounting domain: The case for auditing. *Intelligent Systems in Accounting, Finance & Management: International Journal, 14*(3), 77–86.

Borins, S. (2001). Innovation, success and failure in public management research: Some methodological reflections. *Public Management Review, 3*(1), 3–17.

Boyd, M., & Wilson, N. (2017). Rapid developments in artificial intelligence: How might the New Zealand government respond? *Policy Quarterly, 13*(4), 1–84.

Bracci, E., & Vagnoni, E., (2006). Managerialism, accounting information system and accountability in the Italian local governments: an empirical analysis. In *4th International conference on Accounting, Auditing and Management in Public Sector Reforms*. EIASM.

Cheng, M. Y., & Roy, A. F. (2011). Evolutionary fuzzy decision model for cash flow prediction using time-dependent support vector machines. *International Journal of Project Management, 29*(1), 56–65.

Chui, K. T., Lytras, M. D., & Visvizi, A. (2018). Energy sustainability in smart cities: Artificial intelligence, smart monitoring, and optimization of energy consumption. *Energies, 11*(11), 1–20.

Cohen, J., Krishnamoorthy, G., & Wright, A. M. (2002). Corporate governance and the audit process. *Contemporary Accounting Research, 19*(4), 573–594.

Ciasullo, M. V., Fenza, G., Loia, V., Orciuoli, F., Troisi, O., & Herrera-Viedma, E. (2018). Business process outsourcing enhanced by fuzzy linguistic consensus model. *Applied Soft Computing, 64*, 436–444.

De Maio, C., Botti, A., Fenza, G., Loia, V., Tommasetti, A., Troisi, O., & Vesci, M. (2015, November). What-if analysis combining fuzzy cognitive map and structural equation modeling. In *2015 Conference on Technologies and Applications of Artificial Intelligence (TAAI)* (pp. 89–96). IEEE.

D'aniello, G., Gaeta, A., Gaeta, M., Lepore, M., Orciuoli, F., & Troisi, O. (2016). A new DSS based on situation awareness for smart commerce environments. *Journal of Ambient Intelligence and Humanized Computing, 7*(1), 47–61.

de Sousa, W.G., de Melo, E.R.P., Bermejo, P.H.D.S., Farias, R.A.S. & Gomes, A.O., (2019). How and where is artificial intelligence in the public sector going? A literature review and research agenda. *Government Information Quarterly, 36*(4), 101392.

Desouza, K. C., Dawson, G. S., & Chenok, D. (2020). Designing, developing, and deploying artificial intelligence systems: Lessons from and for the public sector. *Business Horizons, 63*(2), 205–213.

Elliott, R. K. (1992). The third wave breaks on the shores of accounting. *Accounting Horizons, 6*(2), 61–85.

Guba, E. G., & Lincoln, Y. S. (2005). Paradigmatic controversies, contradictions, and emerging confluences. *The Landscape of Qualitative Research*, 255–286.

Hennink, M., Hutter, I., & Bailey, A. (2020). *Qualitative research methods*. SAGE Publications Limited.

Horton, J., Macve, R., & Struyven, G. (2004). Qualitative research: experiences in using semi-structured interviews. In *The real life guide to accounting research* (pp. 339–357). Elsevier.

Issa, H., Sun, T., & Vasarhelyi, M. A. (2016). Research ideas for artificial intelligence in auditing: The formalization of audit and workforce supplementation. *Journal of Emerging Technologies in Accounting, 13*(2), 1–20.

Kend, M., & Nguyen, L. A. (2020). Big Data analytics and other emerging technologies: The impact on the Australian audit and assurance profession. *Australian Accounting Review, 90*(1), 1–14.

Kokina, J., & Davenport, T. H. (2017). The emergence of artificial intelligence: How automation is changing auditing. *Journal of Emerging Technologies in Accounting, 14*(1), 115–122.

Mbecke, Z. M. P. (2014). Resolving the service delivery dilemma in South Africa through a cohesive service delivery theory. *Problems and Perspectives in Management, 12*(4), 265–275.

Metaxiotis, K., & Psarras, J. (2003). Expert systems in business: Applications and future directions for the operations researcher. *Industrial Management & Data Systems, 103*(1), 361–368.

Mikhaylov, S. J., Esteve, M., & Campion, A. (2018). Artificial intelligence for the public sector: Opportunities and challenges of cross-sector collaboration. *Philosophical Transactions of the Royal Society a: Mathematical, Physical and Engineering Sciences, 376*(2128), 1–21.

Moll, J., & Yigitbasioglu, O. (2019). The role of internet-related technologies in shaping the work of accountants: New directions for accounting research. *The British Accounting Review, 51*(6), 1–20.

Mosteanu, N. R., & Faccia, A. (2020). Digital Systems and new challenges of financial management-fintech, XBRL blockchain and cryptocurrencies. *Quality-Access to Success, 21*(174), 159–166.

Potts, J. (2009). The deficit of innovation in public sector: The curious case of too much efficiency and not enough waste and failure Innovation. *Management Policy and Practice, 1*(11), 34–43.

Polese, F., Troisi, O., Torre, C., & Maione, G. (2017). Performance evaluation and measurement in public organizations: A systematic literature review. *International Journal of Business Administration, 8*(1), 106–117.

Qu, S. Q., & Dumay, J. (2011). The qualitative research interview. *Qualitative Research in Accounting & Management, 8*(3), 238–264.

Rogers, E. M. (2002). Diffusion of preventive innovations. *Addictive Behaviors, 27*(6), 989–993.

Rogers, E. M. (2010). *Diffusion of innovations*. Simon and Schuster.

Salawu, M. K., & Moloi, T. S. (2020). Critical Factors For Accounting estimation of investment in artificial intelligence: an imperative for accounting standards setters in the fourth industrial revolution Era. *The Journal of Accounting and Management, 10*(1), 39–48.

Sánchez, S. N. (2019). The implementation of decentralised ledger technologies for public procurement: blockchain based smart public contracts. *Eur. Procurement & Pub. Private Partnership L. Rev., 14*(1), pp.180–196.

Silverman, D. (Ed.). (2020). *Qualitative research.* Sage.

Sun, T. Q., & Medaglia, R. (2019). Mapping the challenges of artificial Intelligence in the public sector: Evidence from public healthcare. *Government Information Quarterly, 36*(2), 368–383.

Sutton, S. G., Holt, M., & Arnold, V. (2016). "The reports of my death are greatly exaggerated"—Artificial intelligence research in accounting. *International Journal of Accounting Information Systems, 22,* 60–73.

Troisi, O., Maione, G., Grimaldi, M., & Loia, F. (2020). Growth hacking: Insights on data-driven decision-making from three firms. *Industrial Marketing Management, 90*(1), 538–557.

Troisi, O., Visvizi, A., & Grimaldi, M. (2021). The different shades of innovation emergence in smart service systems: The case of Italian cluster for aerospace technology. *Journal of Business & Industrial Marketing.* https://doi.org/10.1108/JBIM-02-2020-0091

Ukpong, E. G., Udoh, I. I., & Essien, I. T. (2019). Artificial intelligence: opportunities, issues and applications in banking, accounting, and auditing in Nigeria. *Asian Journal of Economics, Business and Accounting,* 1–6.

Ven, A., & Rogers, E. M. (1988). Innovations and organizations: Critical perspectives. *Communication Research, 15*(5), 632–651.

Visvizi, A., & Lytras, M. (Eds.). (2019). *Smart Cities: Issues and Challenges: Mapping Political.* Elsevier.

Visvizi, A., & Lytras, M. D. (Eds.). (2019). *Politics and technology in the post-truth era.* Emerald Publishing.

Visvizi, A., & Lytras, M. D. (2018). Rescaling and refocusing smart cities research: From mega cities to smart villages. *Journal of Science and Technology Policy Management, 9*(2), 134–145.

Visvizi, A., Lytras, M. D., & Daniela, L. (2018b). Education, innovation and the prospect of sustainable growth and development. In *The future of innovation and technology in education: Policies and practices for teaching and learning excellence.* Emerald Publishing Limited.

Visvizi, A., Lytras, M. D., Damiani, E., & Mathkour, H. (2018). Policy making for smart cities: Innovation and social inclusive economic growth for sustainability. *Journal of Science and Technology Policy Management, 9*(2), 126–133.

Wirtz, B. W., Weyerer, J. C., & Geyer, C. (2019). Artificial intelligence and the public sector—applications and challenges. *International Journal of Public Administration, 42*(7), 596–615.

Zemánková, A. (2019). Artificial intelligence and blockchain in audit and accounting: Literature review. *WSEAS Transactions on Business and Economics, 16*(1), 568–581.

Gennaro Maione PhD in economics and management of public companies, Research Fellow at the University of Salerno, Guest Editor of the journal Sustainability, Topic Editor of the journal Systems, and Editorial Board Member of Open Journal of Accounting, Journal of Business Administration Research, Journal of Business and Economics, American Journal of Accounting, and American Journal of Economics. Research areas: innovation accounting; accounting ethics; accountability of public administrations. Email: gmaione@unisa.it ORCID: 0000-0002-9167-6369

Giulia Leoni Ph.D. candidate in Management and Law – Curriculum Business and Administration, Dept. of Management, Polytechnic University of Marche (Ancona). Research areas: Performance Management Systems; Accounting innovation; Accountability in public sector. Email: g.leoni@pm.univpm.it ORCID: 0000-0003-3932-0901

Artificial Intelligence and Local Governments: The Case of Strategic Performance Management Systems and Accountability

Giulia Leoni⑩, Francesco Bergamaschi⑩, and Gennaro Maione⑩

Abstract Technological advances (from almost unlimited computer power to Artificial Intelligence-based systems) combined with managerial information systems may prove significantly valuable under two different areas. On the one hand, Business Intelligence and Business Analytics, the latter of which is a layer on which Artificial Intelligence commonly builds, may overcome an information overload, absence of cause-effect relationships and a lack of holistic views of the organization. On the other hand, they may improve accountability while also engaging different stakeholders. However, the diffusion of this approach, especially in the public sector and at a local level, is still both theoretically and practically in its infancy. Thus, this conceptual paper highlights the pivotal role of data-driven decision-making and data-visualization in Local Governments, involving the points of view of both accounting scholars and an Artificial Intelligence practitioner. The literature review suggests how data analytics may allow the alignment between political programs, strategic goals and their implementation, fostering the integration between strategic and operational goals. Moreover, data analytics may increase accountability and support the managerial and political decision-making process, increasing their awareness. This paper calls for new models for public administration managerial decision-making, reporting, and organizational culture which encompass analytics, capable of simplifying data and amplifying its value.

Keywords Artificial intelligence (AI) · Business intelligence · Business analytics · Strategic performance management systems · Accountability · Local governments

G. Leoni
Department of Management, Polytechnic University of Marche, Ancona, Italy
e-mail: g.leoni@pm.univpm.it

F. Bergamaschi
Department of Management, University of Bologna, Forlì, Italy
e-mail: francesco.bergamaschi@unibo.it

G. Maione (✉)
Department of Economics and Statistics (DISES), University of Salerno, Fisciano, Italy
e-mail: gmaione@unisa.it

© The Author(s), under exclusive license to Springer Nature Switzerland AG 2021
A. Visvizi and M. Bodziany (eds.), *Artificial Intelligence and Its Contexts*, Advanced Sciences and Technologies for Security Applications,
https://doi.org/10.1007/978-3-030-88972-2_10

1 Introduction

Big data availability and technological advances have led the combination of Business Intelligence (BI), Business Analytics (BA) and Artificial Intelligence (AI) with managerial information systems, identifying them as a solution for diagnostic issues related to strategic performance management systems (SPMS) (Polese et al., 2017; Silvi et al., 2015) and business accountability, englobing the economic, social and environmental sustainable dimensions (Kaplan & McMillan, 2020; Lytras et al., 2020; Mazzara et al., 2010;). The demand for accountability and performance outcome improvement of LGs is growing and is characterized by multiple goals to achieve on behalf of diverse stakeholders (Wisniewski & Olafsson, 2004). Even when LGs are networked with the joint-delivery of services, though, they often adopt ineffective and still insular SPMS (Minassians, 2015). Despite recent extensive studies in the private sector and the high benefits obtained by a combination between BI, BA and AI with SPMS, this concept seems both theoretically and practically in its infancy in the public sector and at a local level. Yet, knowledge and experience about intelligent IT systems is not so common (Baldwin et al., 2006). This paper provides public-sector contextualized knowledge of Artificial Intelligence in SPMS and accountability processes, involving the points of view both of accounting scholars and an AI practitioner. The main source of our research to identify AI potentiality for LG SPMS consisted of academic and management publications, discussing both theoretical framework and real-world experience of LGs with AI. The paper starts with an outline of the prior literature on SPMS and accountability (Sect. 2). This literature review is intended to provide a background and context to the research, rather than develop specific theories and hypothesis testing. A short discussion of SPMS and accountability in LGs are subsequently provided (Sect. 3). The paper concludes with a discussion of the main contributions of the research and possible consequences for using technological advances as a tool to support SPMS and accountability in LGs (Sect. 4). Emerging issues and suggestions for further research are then presented (Sect. 5).

2 Strategic Performance Management Systems and Accountability

SPMS literature (Silvi et al., 2015; Melnyk et al., 2014; de Waal, 2007) investigates operational and strategy-oriented approaches to the discipline, and primarily focuses on multidimensional perspectives, adopting the concept of sustainability in its triple form: economic, social and environmental (Kaplan & McMillan, 2020; Lytras et al., 2020; Mazzara et al., 2010;). According to de Waal (2007), SPMS is a managerial and analytical process "[...] where steering of the organization takes place through the systematic definition of mission, strategy and objectives of the organization, making these measurable through critical success factors and key performance indicators

(KPIs), in order to be able to take corrective actions to keep the organization on track". Thus, decision-makers use SPMS to pursue the mission, strategy and objectives of an enterprise, so that operational strategies and performance measurement strive in the same direction. Following the emergence of research on triple-bottom line in accounting and reporting, it has been understood that even businesses need more than traditional financial information about their past performance to manage themselves conscientiously and to provide accountability for investments that improve environmental and societal outcomes (Kaplan & McMillan, 2020; Smith, 2005). In particular, it has been highlighted by several authors (Kaplan & McMillan, 2020; Silvi et al., 2015; Kaplan & Norton, 1992) how SPMS integrates financial measures with non-financial indicators and considers an internal and external vision, emphasizing a forward-looking perspective and the definition of cause-effect relationships among diverse measures and perspectives included in the system.

However, it has been documented that managerial accounting techniques developed by academics and practitioners (i.e. Kaplan & Norton, 1992), conceived as knowledge-intensive tools, have arisen a gap between theory and practice. Indeed, despite the application of multidimensional models, the numerous dimensions are analysed with conventional metrics, and businesses continue to prevalently focus on the traditional short-term financial perspective (Silvi et al., 2015; Smith, 2005). Therefore, they are internally oriented and backward-looking, lacking the cause-effect relationships between different dimensions. Among SPMS, the Balanced Scorecard (BSC) (Kaplan & Norton, 1992) has received major attention from managerial accounting scholars, also analysing the public sector (Bobe et al., 2017; Hoque, 2014) and the LG context (Northcott & Taulapapa, 2012; Wisniewski & Olafsson, 2004). Although the valuable diverse SPMS contribution ranged from private to public sector and from single organization to collaborative networks, a common reasoning line to be implemented does not seem possible (Graça & Camarinha-Matos, 2017). Moreover, the desirability to implement SPMS also in collaborative networks appears to be unrelated with SPMS usefulness and effectiveness (Duan & Park, 2010).

Effective SPMS should reduce the logical and operative distance between strategic development and decision-making at the governance level and at the project management level (Del Bene, 2009), however, instances of misalignment are still prevalent (Melnyk et al., 2014). Moreover, SPMS should try to monitor emerging strategies on an ongoing basis through its process of learning (Simons, 1994). SPMS can be defined as effective if any specific strategic factor (i.e. quality) is aligned with its importance, thus, providing a high level of information on the most important strategic factor and a low level for the marginal ones. However, higher focus on differentiation is not followed by greater information effectiveness on strategic differentiation targets (Silvi et al., 2015). To address this challenge, SPMS articles (Appelbaum et al., 2017; Raffoni et al., 2018), derived from the private sector, consider the potentiality of technological advances on managerial accounting systems. In particular, a positive influence has been highlighted in two different areas. On the one hand, BI and BA, the latter of which is a layer on which Artificial Intelligence commonly builds, may overcome an information overload, absence of cause-effect relationships and a lack of holistic views of the organization (Holsapple et al., 2014). On the other hand, they may improve the required 'broadened' accountability while also engaging different stakeholders (Visvizi & Lytras, 2019a; Visvizi et al., 2018b). Prior studies, that also take into consideration LG levels, have constantly highlighted the success of SPMS use when the performance information influence decision-making and accountability, defined by the literature as 'instrumental use' (Julnes, 2008). Given the aforementioned potentiality of advanced technology for SPMS, it seems reasonable to further deepen the understanding on its implementation in the Public sector.

3 Strategic Performance Management Systems and Accountability in LGs

Since the '90 s, public SPMS adoption in public sector organizations has accelerated and performance measurement has expanded both horizontally, in terms of input, activities, output and outcomes, and vertically, in terms of micro, meso and macro levels (Bouckaert & Halligan, 2008). LGs, even if increasingly networked with the joint-delivery of services, often adopt ineffective and still insular SPMS (Minassians, 2015). One of the main challenges is the design of performance indicators that reflect multilevel performance of multisectoral priorities for management (Provan et al., 2007): target and benchmark often missing, output and input are used most often together with a focus on procedures, while productivity indicators, quality, and cost measurement are still notoriously absent. Moreover, the selection of a manageable number of appropriate KPIs and the mapping of casual links among performance measures, when implementing SPMS, seems more problematic in the complex public sector context– as with LGs, often characterized by a multiplicity of activities, output and objectives with the aim to solve problems and enhance welfare (outcomes) (Northcott & Taulapapa, 2012). Thus, SPMS often appear to be in place but without supporting the decision-making or accountability processes (Aleksandrov et al., 2020), concluding that result-based management is still unsuccessful.

Studies (amongst others e.g. Aleksandrov et al., 2020; Northcott & Taulapapa, 2012; Van de Walle & Van Dooren, 2008) show that, among the reasons for SPMS implementation within the public sector, performance measuring and reporting requirements seem the most recurrent. They are not always aligned with the actual organizational needs and the learning-based approach, which is essential to effectively implement and use SPMS. Prior studies (e.g. Northcott & Taulapapa, 2012) suggest that performance causal relationships may be difficult especially in LG contexts, where managers feel that their organization lacks a clear strategic direction and consider organizational outcomes as heavily influenced by external forces beyond their control (e.g. legislation, policy and politics). The importance of strategic planning has been highlighted (e.g. Andrews et al., 2012; Johnsen, 2019) in order to achieve a realistic and sustainable SPMS pathway aimed to establish a 'dashboard' of measures capable of describing organizational objectives and of encouraging coherent behaviours based on the integration of financial and non-financial indicators (Silvi et al., 2015). However, there still is a lack of a perceived strategic orientation within public sector organizations (Mazzara et al., 2010; Northcott & Taulapapa, 2012), which may impede SPMS adoption since its effective use is predicated on an assumption that the organization is strategy driven. This suggests that SPMS in LGs may not follow an adequate methodological path that, on the basis of the objectives, identifies the critical factors to be monitored (Del Bene, 2009).

However, even where the strategy was implemented, also adopting the increasingly required sustainability concept, there seemed to be a lot of frustration about the lack in the SPMS design (Brorström et al., 2018). Moreover, as the literature suggests, the possible SPMS tension will be highlighted by the need for LG "performance culture" rather than "measurement culture" (Van de Walle & Van Dooren, 2008). This points out that analytical performance measurement as a system to ensure a focus on results and accountability is still a cultural problem. Indeed, SPMS in LGs are often complex, due to an information overload which may increase resistance to the use of performance information (Van de Walle & Van Dooren, 2008; Van Thiel & Leeuw, 2002). Moreover, the practice of SPMS appears difficult when moving from strategy and the objective formulation to the operationalization of these objectives, through the development of a performance measurement system (Brorström et al., 2018; Van Thiel & Leeuw, 2002).

Thus, SPMS seems to be adopted as an important mechanism for external accountability, distracting from internal accountability demanded by politicians and citizens in LGs. However, a critical analysis on the significant activities to be undertaken to achieve performance targets and outcomes and to give accounts of future action should be required for the usefulness of SPMS. To this extent, the law does not entail a managerial action directed at improving outcomes on the "leading indicators" to drive performance improvement in the "lagging dimension". This may lead to unintended consequences which may not only invalidate conclusions on the performance itself but may also "create dysfunctional effects such as ossification, a lack of innovation, tunnel vision, and suboptimization" (Van Thiel & Leeuw, 2002, p.270). Therefore, SPMS, where adopted, lead to some skepticism about their usability and feasibility in the public context and, therefore, to the misuse of information (Ciasullo et al., 2020; Troisi et al., 2021). The potential of SPMS is undermined and there is a clear need to research how to sustain public decision-makers' engagement. There is a growing body of evidence outlining the pivotal role of the impact of technological advances on SPMS, but little is known within the public sector and at a local level (Reis et al., 2019). Considering the absence of advanced technological techniques and the extreme scarcity of structured SPMS processes, it seems important to highlight the remarkable AI-based system role in SPMS and to take into consideration the BI and BA concepts. Thus, the following paragraph aims to explain the potentiality and challenges of AI in terms of LG decision-making processes and accountability, based on literature and insights from AI.

4 Artificial Intelligence as a Tool to Support Strategic Performance Management Systems and Accountability in LGs

Given the LG constraints for SPMS implementation and the increasing information needs coming from uplevel government but also from politicians, managers and citizens in general, the question arises of how LGs should deal with the AI potentiality and the challenges involved. With technological advances, from almost unlimited computer power to AI-based systems, the importance of data at a global level is remarkable: between 2018 and 2025, 530% increase of global data volume, 10,9 million data professionals in the EU (5,7 million in 2018) and 829 b€ value of data market in the EU (301 b€ in 2018) are forecasted (European Commission, 2019). The data landscape involves three main areas of use where SPMS could rely upon: BI, BA and AI. BI and BA belong to the family of *Decision Support Systems* (DSS). The term *Business Intelligence* was first used in 1958 by IBM and refers to the handling of *structured* data (e.g. accounting data stored in tables inside databases), to create reports able to support the management decision-making process timing and quality. Structured data stands in contrast to *non-structured data* (e.g. not stored in database tables, e.g., videos, texts, comments and tags). The horizon of the BI expanded naturally into BA, heavily including Statistics applied on massive datasets (Holsapple et al., 2014). The purpose of this was to extend the analysis not only on the facts *happened*, but also to generate insights on what was not a *fact* yet.

BI purpose is to deal with the connection to different kind of data sources and not only with the cleansing, reshaping, selection and aggregation of the collected data—all these phases known collectively as ETL (Extract, Transform and Load) process, but also with its visualization and communication. The connection gathers not only data but also *metadata* (the structure of data into tables like column names, column data types, data distribution), arranged in a *data model*, in such a way as to generate timely available, relevant and accurate KPIs to manage the business at hand, dynamically calculating them at different level of granularities to provide insights to the entire organization, from operational to management levels (Provan et al., 2007). As of today, most of corporate departments are having interest in BI (Kohtamäki & Farmer, 2017). Furthermore, Watson (2009) shows how BI is not only to be considered of interest within the organization but also for its suppliers, customers and regulators, an approach known as *Pervasive* BI, that points to a collaborative network creation. BI output is a cleansed, stable and coherent *semantic model* describing what *has happened* in an organization and able to generate simple forecasts and scenario analysis. Its output is the ideal input for a BA process, which is highly sensitive to the quality of the input data. BA systems, in fact, take advantage of the ETL process that is part of BI and cleans the data from noise and various pollutions. BA purpose is to help the acquisition of insights that help the decision-making process like estimations of problem-solving drivers, cause-effect relationships and scenarios likelihood. BI and BA are often fully integrated into DSS systems, creating a full flow from the original data to scenario likelihood estimations. SPMS could utilize BA to answer questions as: (1) "what has happened?", "what is happening?" (Descriptive Analytics) (2) "what will happen?" (Predictive Analytics) (3) "what is the optimized solution?" (Prescriptive Analytics) (Mortenson et al., 2014).

BA has gone through several eras, with the current one commonly referred to as *Analytics 4.0*, in which AI is deeply involved with autonomous Machine Learning systems (Davenport, 2018). Artificial Intelligence (AI) deals with systems able to apply BA in an autonomous way (Arasteh et al., 2016; Ciasullo et al., 2018), allows the automatic learning through, for example, *Artificial Neural Networks* (ANN) and its applications ranges from Marketing (Kohtamäki & Farmer, 2017; Watson, 2009) to Security (Costantino, 2020). ANN can be broadly defined as a massive combination of simple parallel processing units, which can acquire knowledge from the environment, through a learning process and store the knowledge in its connections (Guresen and Kayakutlu, 2011). In this context, algorithms adapt themselves to the new data, self-adjusting their decision path effectively, not executing instructions passively but rather finding patterns in the data, to turn them into computational steps (Kouziokas, 2017). Companies AI initiatives should be considered as a natural extension of BI and BA models running. AI systems rely, in fact, often—but not always—on statistical assumptions on data and statistical methods, which ties them to the BA process as a foundation layer. *Machine learning (ML)* and *Deep learning (DL)* techniques are examples of AI processes. ML models need to be *trained*. For example, a ML model attempting to predict the missing payments would need to be designed on a dataset in which that phenomenon has been observed. The resulting model is *tested* with a *validation* dataset, and the outcomes predicted by the model are compared to the (known) outcomes. If this test is passed, the model is used to *predict* values for which the outcome is not known (Davenport, 2018). DL models are an advanced form of artificial neural networks that can *train other networks*. The trained networks are then used to perform different tasks.

Typical applications of DL are automatic voice and image recognition (Davenport, 2018). Both BA and AI systems leverage large datasets and employ statistical methods to unlock new sources of value, insights and problem-solving approaches, both in a quantitative and qualitative way. Organizations having BA systems in place can then benefit from them to transition into AI systems, as the easiest path to a successful transition to those systems is through the extension of pre-existing BA systems (Davenport, 2018). Public administrations have adopted AI systems to deal with, among other applications of public interests, security and crime prevention. Kouziokas (2017) shows how Artificial intelligence was used in Chicago to build artificial neural network predictive models, to predict the crime risk efficiently and effectively in different urban areas. Time Series predictions and spatial clustering were the key technology used, allowing the public administration to adopt effective planning and intervention strategies to ensure safety in transports. There are examples of AI systems with the purpose of crime prevention in several Italian LGs as well (Costantino, 2020). Raffoni et al. (2018) highlights how BA interacts with SPMS, defining *Business Performance Analytics* (BPA) as "*the control of business dynamics and performance through the systemic use of data and analytics methods*" and shows that, even if the BA impact on performance management systems is still in an early stage, there are some early studies that show its potential. Furthermore, BPA is found to potentially be able to support SPMS, identifying the most influential variables on performance and to estimate cause-effect relationships. Raffoni et al. (2018) identifies BPA implementation challenges as well: data quality and related competences and the need for a cultural shift to benefit from BPA. These challenges appear especially intense for LGs, due to the generically low education of the staff on BI and BA and the low spread of a data-driven culture (Costantino, 2020). BPA is considered highly valuable since it might help the transition to SPMS that involves target and benchmarks, the calculation of slope coefficients and risk analysis (Raffoni et al., 2018). Costantino (2020) shows how Italian public administration finds itself in front of a challenge: digitalization of the PA in Italy has been deployed more effectively where officials average age was significantly lower than the average. The staff is, anyway, overall high-aged and its education is not in line with the modern technologies used in digitalization processes. It is critical, therefore, to review the staff training approach, which has been historically focused on economics and law, not on computational technologies. Also, it is critical to assure the entry of young and highly educated scientists that can deal immediately with these technologies. Davenport (2018) shows that the availability of open-source AI services, provided by many web services players, can speed up the adoption of BI/BA/AI technologies. This makes AI services like image recognition and natural language processing readily available to LGs.

At the same time, AI processes automate what in BA has to be implemented with human intense activity, so implementing the AI services available on existing BI/BA processes can free up the time of the new and young scientists, allowing them to focus on high-level strategic projects. AI can then help shortening the time needed to deploy new BI/BA processes into the organizations or to improve the existing ones. Raffoni et al. (2018) show that there is a relationship between the use of BA systems and the adoption of advanced SPMS: BA systems can in fact act as a complement to SPMS, identifying and quantifying critical drivers to performance. This suggests that the transition of LGs to the adoption of SPMS, both operational and strategic, might be facilitated using BA systems. Costantino (2020), shows that a critical obstacle to the adoption of BA and AI systems in the public sector is the unavailability of trained staff in order to design, use, understand the output of BA systems. A further step, going from BA to AI systems, might facilitate—thanks to the autonomous application—BA system implementation where the knowledge needed is unavailable (Davenport, 2018). In addition, AI can help especially in LGs, where the organizational focus is spread among several different aspects of society and the creation, maintenance and update of relevant KPIs and their drivers is especially problematic (Northcott & Taulapapa, 2012).

Finally, AI systems could help using SPMS more for internal accountability, as their insights are not subject to the LGs legal reporting obligations nor to any political pressure, and to drive attention to significant cause-effect relationships influencing the performance, having the chance to bring managers up to speed on a continuous basis and avoiding information overload, focusing on what has a real impact on performance on various LG dimensions. This might increase LG awareness of the potential of SPMS, beyond pure reporting. To maximize the probabilities of success in the adoption of AI processes, it is critical for organizations to start by honestly assessing their technology and data infrastructures, to understand the technology gap between current capabilities and the capabilities needed to reach the ability to use AI systems. Davenport (2018) lists four areas to start the assessment: company culture, existing analytics teams' capabilities, data and technology platforms capabilities and individual persons' capabilities and suggests the creation of a center of excellence (COE) that can act as internal partner to other development and analytics teams. COEs should also ensure that communication with all the key stakeholders is kept active.

5 By Means of Conclusion

AI application to SPMS is currently heavily discussed in scientific literature but mainly focused on the private sector and is still more focused on Business Analytics (i.e. Raffoni et al., 2018). Therefore, the literature is largely silent on the challenges and potentiality of AI in the public sector and, specifically, those of LGs characterized by specific peculiarities which should be taken into consideration for seeking ways to improve SPMS implementation. This paper aims to give a first conceptual overview providing LG contextualized knowledge of AI in SPMS, and accountability processes. In particular, this paper provides several insights on how data analytics and AI may allow the alignment between political programs, strategic goals and their implementation, fostering the integration between strategic and operational goals. Firstly, since its autonomous application, AI may overcome the criticism related to the unavailability of trained staff in LGs (Costantino, 2020). The findings emerged from the analysis suggested that the adoption of AI may boost advanced and effective SPMS focused on a learning-based approach and capable of selecting a manageable number of KPIs as well as mapping casual links among performance measures, therefore grabbing the actual organizational needs. This view can be coherent with the Raffoni et al. (2018) study where the relationship between the use of BA systems and the adoption of advanced SPMS has been found. This could be a key issue for LGs, where there is a low spread of data-driven culture (Costantino, 2020) and where SPMS is often implemented for performance measuring and reporting requirements, which, in most cases, are not aligned with local needs (Aleksandrov et al., 2019). Therefore, AI would help the objective and learning-process for the production of performance metrics potentially prompting the information use of public management and democratic accountability (Julnes, 2008). Moreover, AI application, through the identification of causal relationships between performance measures, could potentially fill the gap between strategic development and decision-making at the governance level and at the project management level. In this regard, it would avoid information overload, and approach a "performance culture" (Hood, 2006), making operationalization of sustainable strategies possible (Brorström et al., 2018). Due to the lack of knowledge about these topics, all the aforementioned suggestions could be empirically analyzed by future research. This further analysis would also require to understand the pivotal role of smart cities (Lytras et al., 2020; Visvizi & Lytras, 2018a) together with the need of new managerial paradigm for decision and policy making processes in the public sector (Visvizi & Lytras, 2019b; Visvizi et al., 2018c).

References

Aleksandrov, E., Bourmistrov, A., & Grossi, G. (2020). Performance budgeting as a "creative distraction" of accountability relations in one Russian municipality. *Journal of Accounting in Emerging Economies, 10*(3), 399–424. https://doi.org/10.1108/JAEE-08-2019-0164.

Andrews, R., Boyne, G., Law, J., & Walker, R. (2012). *Strategic Management and Public Service Performance*. Palgrave MacMillan. https://doi.org/10.1057/9780230349438.

Appelbaum, D., Kogan, A., Vasarhelyi, M., & Yan, Z. (2017). Impact of business analytics and enterprise systems on managerial accounting. *International Journal of Accounting Information Systems, 25*, 29–44. https://doi.org/10.1016/j.accinf.2017.03.003.

Arasteh, H., Hosseinnezhad, V., Loia, V., Tommasetti, A., Troisi, O., Shafie-khah, M., & Siano, P. (2016). Iot-based smart cities: A survey. In *2016 IEEE 16th International Conference on Environment and Electrical Engineering (EEEIC)* (pp. 1–6). IEEE. https://doi.org/10.1109/EEEIC.2016.7555867.

Baldwin, A. A., Brown, C. E., & Trinkle, B. S. (2006). Opportunities for artificial intelligence development in the accounting domain: The case for auditing. *Intelligent Systems in Accounting, Finance & Management: International Journal, 14*(3), 77–86. https://doi.org/10.1002/isaf.277.

Bobe, B. J., Mihret, D. G., & Obo, D. D. (2017). Public-sector reforms and balanced scorecard adoption: An Ethiopian case study. *Accounting, Auditing and Accountability Journal, 30*(6), 1230–1256. https://doi.org/10.1108/AAAJ-03-2016-2484.

Bouckaert, G., & Halligan, J. (2008). Comparing performance across public sectors. In *Performance Information in the Public Sector* (pp. 72–93). Palgrave Macmillan. https://doi.org/10.1007/978-1-137-10541-7.

Brorström, S., Argento, D., Grossi, G., Thomasson, A., & Almqvist, R. (2018). Translating sustainable and smart city strategies into performance measurement systems. *Public Money & Management, 38*(3), 193–202. https://doi.org/10.1080/09540962.2018.1434339.

Ciasullo, M. V., Fenza, G., Loia, V., Orciuoli, F., Troisi, O., & Herrera-Viedma, E. (2018). Business process outsourcing enhanced by fuzzy linguistic consensus model. *Applied Soft Computing, 64*, 436–444. https://doi.org/10.1016/j.asoc.2017.12.020.

Ciasullo, M. V., Troisi, O., Grimaldi, M., & Leone, D. (2020). Multi-level governance for sustainable innovation in smart communities: An ecosystems approach. *International Entrepreneurship and Management Journal, 16*(4), 1167–1195. https://doi.org/10.1007/s11365-020-00641-6.Iniziomodulo.

Costantino, F, (2020). Public officials and the design of algorithms. Lessons from the Italian experience. *Previdenza Sociale, 875*. https://doi.org/10.4399/978882553896013.

Davenport, T. (2018). From analytics to artificial intelligence. *Journal of Business Analytics, 1*(2), 73–80. https://doi.org/10.1080/2573234X.2018.1543535.

de Waal, A. A. (2007). Strategic performance management. In *A Managerial and Behavioural Approach*. Palgrave MacMillan.

Del Bene, L. (2009). Sistemi informativi per il management pubblico. La contabilità analitica e gli indicatori. In L. Anselmi, F. Donato, A. Pavan, M. Zuccardi Merli (Eds.), *I principi contabili internazionali per le amministrazioni pubbliche italiane, Giuffré, Milano*.

Duan, L. N., & Park, K. H. (2010). Applying the balanced scorecard to collaborative networks. In *2010 6th International Conference on Advanced Information Management and Service (IMS)* (pp. 131–134). IEEE.

European Commission (2019) European Data Strategy 2019–2024. Available at https://ec.europa.eu/info/strategy/priorities-2019-2024/europe-fit-digital-age/european-data-strategy#examples-of-industrial-and-commercial-data-use.

Graça, P., & Camarinha-Matos, L. M. (2017). Performance indicators for collaborative business ecosystems—Literature review and trends. *Technological Forecasting and Social Change, 116*, 237–255. https://doi.org/10.1016/j.techfore.2016.10.012.

Guresen, E., & Kayakutlu, G. (2011). Definition of artificial neural networks with comparison to other networks. *Procedia Computer Science, 3*, 426–433. https://doi.org/10.1016/j.procs.2010.12.071.

Holsapple, C., Lee-Post, A., & Pakath, R. (2014). A unified foundation for business analytics. *Decision Support Systems, 64*(2014), 130–141. https://doi.org/10.1016/j.dss.2014.05.013.

Hoque, Z. (2014). 20 years of studies on the balanced scorecard: Trends, accomplishments, gaps and opportunities for future research. *The British Accounting Review, 46*(1), 33–59. https://doi.org/10.1016/j.bar.2013.10.003.

Johnsen, Å. (2019). Does formal strategic planning matter? An analysis of strategic management and perceived usefulness in Norwegian municipalities. *International Review of Administrative Sciences*, 1–19. https://doi.org/10.1177/0020852319867128.

Julnes, P. L. (2008). Comparing performance across public sectors. In *Performance Information in the Public Sector* (pp. 58–71). Palgrave Macmillan. https://doi.org/10.1007/978-1-137-10541-7.

Kaplan, R. S., & McMillan, D. (2020). Updating the balanced scorecard for triple bottom line strategies. In *Harvard Business School Accounting & Management Unit Working Paper* (pp. 21–28). https://dx.doi.org/https://doi.org/10.2139/ssrn.3682788.

Kaplan, R. S., & Norton, D. P. (1992). The balanced scorecard—measures that drive performance. *Harvard Business Review, 70*(January-February), 71–79.

Kohtamäki, M., & Farmer, D. (2017). *Real-time Strategy and Business Intelligence* (pp. 11–37). Springer International Publishing. https://doi.org/10.1007/978-3-319-54846-3.

Kouziokas, G. N. (2017). The application of artificial intelligence in public administration for forecasting high crime risk transportation areas in urban environment. *Transportation Research Procedia, 24*, 467–473.

Lytras, M. D., Visvizi, A., Chopdar, P. K., Sarirete, A., & Alhalabi, W. (2020). Information management in smart cities: Turning end users' views into multi-item scale development, validation, and policy-making recommendations. *International Journal of Information Management, 56*, 102146. https://doi.org/10.1016/j.ijinfomgt.2020.102146.

Mazzara, L., Sangiorgi, D., & Siboni, B. (2010). Public strategic plans in Italian local governments: A sustainability development focus? *Public Management Review, 12*(4), 493–509. https://doi.org/10.1080/14719037.2010.496264.

Melnyk, S. A., Bititci, U., Platts, K., Tobias, J., & Andersen, B. (2014). Is performance measurement and management fit for the future? *Management Accounting Research, 25*(2), 173–186. https://doi.org/10.1016/j.mar.2013.07.007.

Minassians, H. P. (2015). Network governance and performance measures: Challenges in collaborative design of hybridized environments. *International Review of Public Administration, 20*(4), 335–352. https://doi.org/10.1080/12294659.2015.1088689.

Mortenson, M. J., Doherty, N. F., & Robinson, S. (2014). Operational research from taylorism to terabytes: A research agenda for the analytics age. *European Journal of Operational Research, 241*, 583–595. https://doi.org/10.1016/j.ejor.2014.08.029.

Northcott, D., & Taulapapa, T. M. A. (2012). Using the balanced scorecard to manage performance in public sector organizations: Issues and challenges. *International Journal of Public Sector Management*. https://doi.org/10.1108/09513551211224234.

Polese, F., Troisi, O., Torre, C., & Maione, G. (2017). Performance evaluation and measurement in public organizations: A systematic literature review. *International Journal of Business Administration 8*(1), 106–117. https://doi:https://doi.org/10.5430/ijba.v8n1p106.

Provan, K., Fish, A., & Sydow, J. (2007). Interorganizational networks at the network level: A review of the empirical literature on whole networks. *Journal of Management, 33*(3), 479–516. https://doi.org/10.1177/0149206307302554.

Raffoni, A., Visani, F., Bartolini, M., & Silvi, R. (2018). Business performance analytics: Exploring the potential for performance management systems. *Production Planning & Control, 29*(1), 51–67. https://doi.org/10.1080/09537287.2017.1381887.

Reis, J., Santo, P. E., & Melão, N. (2019). Impacts of artificial intelligence on public administration: A systematic literature review. In *2019 14th iberian conference on Information Systems and Technologies (CISTI)* (pp. 1–7). IEEE. https://doi.org/ https://doi.org/10.23919/CISTI.2019.876 0893.

Silvi, R., Bartolini, M., Raffoni, A., & Visani, F. (2015). The practice of strategic performance measurement systems. *International Journal of Productivity and Performance Management.* https://doi.org/10.1108/IJPPM-01-2014-0010.

Simons, R. (1994). How new top managers use control systems as levers of strategic renewal. *Strategic Management Journal, 15*(3), 169–189. https://doi.org/10.1002/smj.4250150301.

Smith, M. (2005). *Performance measurement and management: A strategic approach to management accounting.* Sage.

Troisi, O., Visvizi, A., & Grimaldi, M. (2021). The different shades of innovation emergence in smart service systems: The case of Italian cluster for aerospace technology. *Journal of Business & Industrial Marketing.* https://doi.org/10.1108/JBIM-02-2020-0091.

Van de Walle, S., & Van Dooren, W. (2008). Introduction: using public sector performance information. In *Performance Information in the Public Sector* (pp. 1–8). Palgrave Macmillan. https://doi.org/10.1007/978-1-137-10541-7_1.

Van Thiel, S., & Leeuw, F. L. (2002). The performance paradox in the public sector. *Public Performance & Management Review, 25*(3), 267–281. https://doi.org/10.1080/15309576.2002.11643661.

Visvizi, A., & Lytras, M. (Eds.). (2019). *Smart cities: Issues and challenges: Mapping political.* Elsevier.

Visvizi, A., & Lytras, M. D. (Eds.). (2019). *Politics and technology in the post-truth era.* Emerald Publishing.

Visvizi, A., & Lytras, M. D. (2018). Rescaling and refocusing smart cities research: From mega cities to smart villages. *Journal of Science and Technology Policy Management, 9*(2), 134–145. https://doi.org/10.1108/JSTPM-02-2018-0020.

Visvizi, A., Lytras, M. D., & Daniela, L., (2018b). Education, innovation and the prospect of sustainable growth and development. In *The future of innovation and technology in education: Policies and practices for teaching and learning excellence.* Emerald Publishing Limited. https://doi.org/10.1108/978-1-78756-555-520181015.

Visvizi, A., Lytras, M. D., Damiani, E., & Mathkour, H. (2018). Policy making for smart cities: Innovation and social inclusive economic growth for sustainability. *Journal of Science and Technology Policy Management, 9*(2), 126–133. https://doi.org/10.1108/JSTPM-07-2018-079.

Watson, H. J. (2009). Tutorial: Business intelligence–past, present, and future. Communications of the Association for Information Systems 25.1:39. https://doi.org/10.17705/1CAIS.02539.

Wisniewski, M., & Olafsson, S. (2004). Developing balanced scorecards in local authorities: A comparison of experience. *International Journal of Productivity and Performance Management, 53*(7), 602–610. https://doi.org/10.1108/17410400410561222.

Giulia Leoni Ph.D. candidate in Management and Law – Curriculum Business and Administration, Dept. of Management, Polytechnic University of Marche (Ancona). Research areas: Performance Management Systems; Accounting innovation; Accountability in public sector. Email: g.leoni@pm.univpm.it ORCID: 0000-0003-3932-0901

Francesco Bergamaschi MEng, MEcon, Adjunct Professor. Research areas: Information Systems, Business Intelligence, Business Analytics, Data Management. Email: francesco. bergamaschi@unibo.it ORCID: 0000-0003-4907-7562

Gennaro Maione Ph.D. in economics and management of public companies, Research Fellow at the University of Salerno, Guest Editor of the journal Sustainability, Topic Editor of the journal Systems, and Editorial Board Member of Open Journal of Accounting, Journal of Business Administration Research, Journal of Business and Economics, American Journal of Accounting, and American Journal of Economics. Research areas: innovation accounting; accounting ethics; accountability of public administrations. Email: gmaione@unisa.it ORCID: 0000-0002-9167-6369

Case Studies

Artificial Intelligence (AI) and ICT-Enhanced Solutions in the Activities of Police Formations in Poland

Jacek Dworzecki⑩ **and Izabela Nowicka**⑩

Abstract Contemporary challenges in ensuring public safety require the use of ICT-enhanced solutions, including artificial intelligence (AI) and related tools and applications. However, regardless of the effectiveness of police operations carried out with the use of this type of ICT-based tools, questions arise as to the impact of these on civil rights and freedoms. Two facets of this issue exist, i.e., on the one hand, very effective tools of police surveillance exist, e.g., control of electronic correspondence, use of information databases on citizens. These tools enable effective detection of crime and their perpetrators. On the other hand, a negligent use of the same tools may violate the individual's constitutional rights and freedoms. The objective of this chapter is to map and assess the ICT-enhanced solutions employed by the Polish police formations in view of the impact of these tools on civil rights and freedoms.

Keywords Artificial intelligence (AI) · ICT · Public safety · Civil rights and freedoms · Poland · Police · Border Guard · State Protection Service

1 Introduction

The functioning of any state is a complex process, which includes, among other, tasks in the field of public safety, implemented both externally, e.g., international police cooperation, and domestically, e.g., police activities conducted in the territory of the country. Public safety is bound with the rise of the modern state and—due to the prerogative of action inherent in 'safety'—constitutes one of the fundamental goals of the state (Visvizi, 2015, p. 30, 2016). The necessity to search for access to information about threats caused by the enemy in the external environment of a country requires the creation of a system securing one's own information needs, especially in the field of security and defence. In Poland, several uniformed formations are responsible for public safety. These include, among other, Police, Municipal Guards, the Internal Security Agency, the Central Anticorruption Bureau, the Road

J. Dworzecki · I. Nowicka (✉)
Military University of Land Forces, Wrocław, Poland
e-mail: Izabela.Nowicka@awl.edu.pl

© The Author(s), under exclusive license to Springer Nature Switzerland AG 2021
A. Visvizi and M. Bodziany (eds.), *Artificial Intelligence and Its Contexts*, Advanced Sciences and Technologies for Security Applications,
https://doi.org/10.1007/978-3-030-88972-2_11

Transport Inspection, the State Protection Service, as well as specialized guards, e.g., the Border Guard, the Hunting Guard, the Railway Security Guard, the Marshal Guard, or the Forest Service. These formations deal, among others, with preventing and combating crimes as well as identifying and detaining offenders. The chapter maps and discusses information and communication technology (ICT) and Artificial Intelligence (AI)-enhanced tools and solutions employed by the Police and other formations, supervised by the Ministry of the Interior and Administration, today. These tools and solutions include biometric identification, machine text translation, data mining, image and speech recognition. Against this backdrop, this the chapter examines how civil rights and freedoms are observed in the framework of police activities with the use of the latest ICT solutions. the argument in this chapter is structured as follows. The following part examines the issue of implementing modern technologies in the activities of the Police. Here the key question addressed is to what extent the efficient use of ICT- and AI-enhanced tools and solutions necessitates compromising civic rights and freedoms. The second part concerns the issue of ICT systems and solutions in the field of data mining, image, or speech recognition and in the field of improving the process of professional training of officers as well as criminal analysis, implemented in the Polish police. In this part three ICT systems used by the Polish police are discussed and modern training tools (emergency vehicle driving simulator) and wearable cameras improving the safety of officers during the intervention. Here the key question that is address is whether these reporting systems improve public safety. As part of the discussion, the positive (in the context of crime detection) and negative (in the context of interference in the rights and freedoms of citizens) aspects of the use of modern ICT solutions by police formations in Poland are elaborated.

2 Restrictions on the Implementation of Modern Technologies in the Police

2.1 Modern Technologies Versus Clauses Restricting Civil Rights and Freedoms

The reality that citizens are deal with today makes perception of constitutional rights and freedoms more flexible. Undoubtedly, people perceive the achievements of modern technologies as a boon. Nevertheless, the disturbing truth comes to us that the implementation of this technology entails not only threats but also encroaches on our rights and freedoms. On the one hand, citizens want to be protected by all possible means against threats to the surrounding world, on the other hand, the same citizens oppose undertaking actions that while necessary to ensure safety and security may have an impact on civil rights and freedoms. Citizens want to have services that can counterbalance not least to the criminal world using the latest technological achievements with their skills, equipment or means. However, what should be emphasized, citizens are, which may seem inconsistent—against it. The Constitution of the Republic of Poland, international agreements ratified by Poland in essence guarantee every human being several rights, such as the right to life, freedom, property or freedom in terms of communication, movement, and in particular, in the context of considerations, the right to respect for the privacy of the individual (RP, 1997). The Constitution regulates the right to privacy, stating in Art. 47 that *Everyone has the right to the legal protection of a private and family life, honor and a good name and to decide about one's personal life. The above statements indicate the obligation of public authorities* to regulate the statutory protection of the right to privacy and to refrain from interfering with private spheres. This provision directly refers to human freedom (Article 31 (1)) and individual freedom (Article 41). In the context of this protection, reference should be made to Art. 49, which in its content indicates that: The freedom and protection of the secret of communication is ensured.

The above rights and freedoms are also guaranteed in acts of international law, including the content of art. 17 of the International Covenant on Civil and Political Rights of 1966 i.e., Section 1 *Nobody shall be subjected to arbitrary or unlawful interference with his or her privacy, family, home, correspondence or to unlawful attacks on his or her honor and reputation* (ICCPR, 1966). Section 2 *Everyone has the right to the protection of the law against such interference or attacks.* The content of article 8 Section 1 of the European Convention on Human Rights of 1950 should be also cited: *Everyone has the right to respect for his private life and his correspondence* (CPHR, 1950).

The above provision clearly indicates that "Restrictions on the exercise of constitutional rights and freedoms may be established only by statue and only if they are necessary in a democratic state for its safety or public order, or for the protection of the environment, health and public morality, or the freedoms and rights of others. These limitations shall not violate the essence of freedoms and rights" (RPO, 2020a, 2020b).

The Constitution of the Republic of Poland, in the wording of Art. 31 Sect. 3, points out that a statutory limitation of the use of freedoms and rights is allowed when they are necessary in a democratic state for the protection of its security or public order, or to protect the natural environment, health or public morals, or the freedom and rights of other persons. The possibility of interference is also included in Art. 8 s. 2 of the European Convention on Human Rights. Its content exhaustively lists the conditions and circumstances justifying the described intrusion by public authorities: *There shall be no interference by a public authority with the exercise of this right except such as is in accordance with the law and is necessary in a democratic society in the interests of national security, public safety or the economic wellbeing of the country for the prevention of disorder or crime, for the protection of health or morals, or for the protection of the rights and freedoms of others* (CPHR, 1950).

The European Court of Human Rights also allows this possibility. Pursuant to Art. 8 of the European Convention on Human Rights, interference by the authorities (legislative, executive, judiciary) in the sphere of this privacy is allowed, provided it meets certain criteria. Any trespass must, however, withstand the test of a three-tier evaluation test. It refers to the source of the restriction, its need and the value for which the restriction is introduced. It should be noted here that the protection of human rights is related to functions such as:

- protecting life and property;
- protecting constitutional guarantees;
- creating and maintaining a feeling of security in the community;
- promoting and preserving civil order and managing public assemblies and events;
- detecting and investigating crime;
- identifying victims and other people in need and ensuring the respect of their rights;
- patrolling and protecting demonstrations;
- securing public places;
- protecting dignitaries, etc. (Grieve et al., 2007, p. 51).

Therefore, there is no doubt that technological changes force interference with constitutionally guaranteed civil rights and freedoms (Alleweldt & Fickenscher, 2018, pp. 10–11). That is why it is so essential to balance the limit of restrictions that can be made so that technological innovations can be used by law, which seems to be an inevitable process.

2.2 Regulatory Frameworks and Technical Innovation in the Police: Issues and Challenges

The issues related to legal and ethical restrictions are discussed in the diverse social, academic and other fora, e.g. the European Commission, the OECD, as well as in the context of business conduct (Langford, 2007, p. 46; Visvizi & Lytras, 2019). In Poland, these matters are dealt with by research centers, universities, governmental and non-governmental organizations (e.g. the Society of Polish Economists). As an example of such an action, one can mention, the project entitled *Legal and criminological aspects of the implementation and use of technologies for the protection of internal security*, carried out in 2011–2012 by a consortium consisting of two scientific and research entities.

The results of the conducted research indicate a relationship between the rapidly progressing technological development and the conscious, orderly and mature participation of entities such as the Police. According to the researcher, "... the enthusiasm for computerization is bipolar, as it may not only support the process of increasing the organization's efficiency, but may also become a threat. In particular, those entities that have not noticed the global innovation trend in time and then have ceased to cope with its progressive nature will be vulnerable to the threat" (Pawłowicz, 2016, pp. 57–68).

The Polish Police seeks to meet the conditions imposed both by the society and the essence of its functioning. Therefore, on its own initiative and under the influence of external factors, it changes the way of operating to that with the use of modern technologies. One of the methods described in the first part of the chapter is to use one's own potential in the implementation of projects allowing for the construction and use of modern technologies.

This creates several organizational and legal problems, as indicated by extensive research into selected ICT projects implemented by police officers and employees (Pawłowicz, 2014, pp. 17–29). The analysis referred to the Command Support System, the ePosterunek application and the Safe User Authentication Mode application. The following conclusions were drawn from the conducted research, indicating the causes of organizational and legal problems in the Police in the context of the activities undertaken. Namely:

- Lack of coherent, based on current legal bases, regulations in the field of ICT projects in the Police,
- Lack of clear, structured rules that are obligatorily applied in the organization would ensure constructive cooperation of the substantive units involved in projects,
- Lack of an effective project management method adapted to the current needs,
- Lack of a clearly defined method of building product requirements, the implementation of which translates into the achievement of strategic goals of the Police,
- Immature culture of the organization, lack of knowledge, will and understanding for the essence and importance of conducted projects,
- Lack of full comprehension of the advantages and benefits of using matrix structures dedicated to the implementation of specific tasks,
- Lack of consistent involvement of the organization's management in implementation of projects resulting from the organization's business goals (Pawłowicz, 2016, pp. 148–152).

Legal problems deriving from the introduction and use of new technologies by the judiciary and law enforcement agencies occurring in individual legal disciplines can be grouped as follows (Pływaczewski, 2017, pp. 29–30):

- constitutional law;
- criminal trial;
- administrative law, including departmental acts;
- financial law.

Obviously, it is not about all the institutions of the above-mentioned branches of law, but only about some, such as the guarantee of fundamental (constitutional) rights and freedoms of the individual (e.g. freedom of correspondence, the right to privacy), respect for the rule of law and the principle of proportionality or the protection of personal data, prohibitions on evidence, fiscal secrets etc.

When it comes to guaranteeing the fundamental (constitutional) rights and freedoms of an individual, one of the most important limitations is a high level of interference with privacy. The need to ensure security (in every sense of the word) or to clarify the matter often involves a violation of the right to the protection of private and family life. The use of modern technologies by, e.g., the police is mainly faced with the lack of legal regulations regarding the use of these technologies and restrictions on the use of data and its openness (Bowling & Sheptycki, 2012, p. 27).

In the reality of the Polish state performance "… for many years, the Spokesperson on Human Rights has been giving alarms about the provisions that allow secret services to gather information about citizens" (RPO, 2020a, 2020b). The situation has worsened since 2016, when the powers of the services were increased. They may, inter alia, obtain data on citizens via permanent Internet connections, without the need to submit applications to telecommunications service providers. Therefore, they can collect data not only when it is necessary to detect the most serious crimes and only when other methods are ineffective, but also when it is simply convenient for the services. Moreover, the services may invigilate foreigners for an indefinite period of time without disclosing this information, even after the inspection has ended. Thus, if the Head of the Internal Security Agency finds that a given person should be under surveillance, his decision will not be verified in any way, neither before nor after the inspection. The necessity to change the principles of surveillance was indicated by the Constitutional Tribunal in July 2014. However, the so-called the surveillance law and the anti-terrorist law increased the powers of the services without introducing a control mechanism. The irregularities were noted, among others, by Venice Commission in the opinion of June 2016. And the obligation to inform the data subject about the access of services to his telecommunications data results from numerous judgments of the ECHR.

Another problematic issue is the protection of the image when entities such as the Police use the monitoring of public places. In this context, the boundary of legal relations runs between constitutional values, such as the right to private life, referred to in Art. 47 of the Polish Constitution and the right to privacy (Art. 51), and the principles that justify their restriction, such as freedom of expression (Art. 54) or freedom of the media (Art. 14).

As already indicated, the Polish Police uses technical innovations in its activities. Therefore, it is obliged to comply with the regulations that adapt to changes in the surrounding reality. An example is, for instance, the use of unmanned aerial vehicles in various aspects of ensuring safety and public order based on national regulations and procedures for the operation of unmanned aerial vehicles, adapted to the Implementing Regulation (EU) 2019/947 of 24 May 2019.

The provisions in force in the Polish Police have also been adapted to the above-mentioned EU implementing regulation, which is reflected in the issue of Order No. 63 of the Police Commander in Chief of 7 October, 2019 on the detailed rules for the use of unmanned aerial vehicles in the Police. The content of this legal act in § 3 indicates that:

- point 1. Police unmanned aerial vehicles are used for the performance of official duties in the Police;
- point 2. The operator or observer of police unmanned aerial vehicles may be a Police employee, with the exception of the performance of statutory Police tasks;
- point 3. The use to support other entities in the implementation of their tasks may take place on the basis of separate contracts and concluded agreements (GDPR, 2016).

3   ICT-Enhanced Tools and Systems in the Polish Police

The police are the largest uniformed formation in Poland. Over 103,000 officers serve there. This service is responsible for the level of public safety and public order.in the country. In its activities, the Polish police uses many methods, techniques and ICT solutions aimed at improving the effectiveness of its activities. Police ICT systems are constructed in such a way as to ensure full protection of data, both obtained by officers, transmitted and stored in police databases. These systems guarantee real-time access for many authorized users. The number of generated data is constantly growing, therefore, in order to obtain it, systems enable their analysis based on built-in reports—reporting systems. On their basis, police officers can quickly and accurately assess the situation and make correct and conscious decisions. These reporting systems have a significant impact on improving safety.

3.1   The National Police Information System

The National Police Information System is the central ICT system of the Polish police, put into operation at the beginning of 2003, the task of which is to provide information services for police officers and employees. It can be stated that the database of the National Police Information System is the core of all ICT systems currently used by the Police. The system interacts with other databases made available to the Police. It operates through the Police Data Transmission Network, which is a separate network without the possibility of access by another network, e.g. external Internet. To gain access to the National Police Information System, a computer connected to the Police Data Transmission Network is required and must be equipped with a security system identifying the user. Depending on the permissions, the user can use data that is collected on one server for all. The National Police Information System stores various types of information, in the form of text, but also multimedia, and all data is available at once. Data search is carried out in many ways, i.e. with the use of different search criteria. Search results are in the form of a record or as data blocks. When searching for data, the user is obliged by the system to follow certain pre-programmed rules. The National Police Information System is equipped with various types of security. It operates in accordance with all Polish and EU regulations regarding the collection and use of data, as well as the security of data storage.

As of 6 April, 2013, new provisions of the Police Commander in Chief regulating the processing of information in the National Police Information System have been in force—Decision No. 125 of the Police Commander in Chief of 5 April, 2013 on the operation of the National Police Information System (GP, 2020).

Pursuant to the regulation, the National Police Information System is a set of data files processed in the ICT systems operated by the Police to perform activities in the field of downloading, obtaining, collecting, checking, processing, and using information, including personal data referred to in art. 20 s. 2a and 2b of the Act on the Police. In the discussed system, information may also be processed, including personal data, which the Police is entitled to collect, obtain and process on the basis of separate acts, if such processing contributes to the coordination of information and more effective organization and implementation of the statutory tasks of the Police in the field of detecting and prosecuting offenders as well as preventing and combating crime with inclusion of protecting human life and health.

The most essential modules of the KSIP system include:

- PERSON—a module for registration and processing of information about persons:

    (1) persons suspected of committing crimes prosecuted ex officio;
    (2) minors committing acts prohibited by law as crimes prosecuted ex officio;
    (3) persons with undetermined identity or trying to hide their identity, as well as information about the corpse of an unknown identity;
    (4) wanted persons:
        (a) hiding from law enforcement and judicial authorities,
        (b) recognized as missing, hereinafter referred to as "missing persons" (DPC. no 125, 2013)

- ITEM—a module designed for registration and processing of information about things, objects, firearms, monuments, related to a crime or an offence and having numbers enabling their unambiguous identification;
- FACT—a module for registration and processing of information about:

    (1) events subject to mandatory reporting to the duty officer of the Police Headquarters and those on duty in voivodship Police headquarters in accordance with the catalogue of types of events specified in the regulations governing the organization of duty service in Police organizational units and about events subject to registration for the purposes of the Bulletin of the National Police Information System;
    (2) offences prosecuted ex officio and committed by minors, prohibited by law as public prosecuted offences, hereinafter referred to as "offences";
    (3) preparatory proceedings in cases of crimes registered in the National Police Information System in the FACT file;
    (4) juvenile delinquency proceedings for acts prohibited by law as prosecuted offences under public prosecution;

(5)   road incidents subject to registration in the Accident and Drivers Records System on the basis of separate provisions on the keeping of road incident statistics;

(6)   statistical data specified in the program of statistical surveys of official statistics for a given year, relating to preparatory proceedings conducted by public prosecutors of common organizational units of the public prosecutor's office and proceedings in cases of minors or criminal offences which are crimes conducted by family judges;

(7)   attempted and carried out suicide attacks and cases of drowning.

- ENTITY—a module for processing procedural information about entities conducting business activity;
- POLICE SEARCH SYSTEM—a module used to process data about wanted persons, things, vehicles and documents, taking into account the databases of Interpol and Schengen Information System;
- POLICE REGISTER OF MASS EVENTS—a module for recording and processing data on organized events, sports events, crisis situations, VIP protection and social events.

The National Police Information System significantly accelerated the work of Police officers and thus the Police units, and radically changed the way they operate. This is also the case with the implementation of mobile devices in the Police with access to this system, the so-called Mobile Wearable Terminals and Mobile Shipping Terminals. At present, the service in the Police is becoming more and more effective and efficient. This has a decisive impact on the public safety in Poland.

3.2   ICT-Enhanced Command Support System in the Police

Following the accession to the European Union, Poland was obliged to establish and launch a uniform ICT Emergency Notification System. The system was to handle notifications to emergency services made to the emergency number 112. In order to ensure fast and reliable data exchange between Provincial Emergency Notification Centers, the Police, fire brigades and ambulance services, it was decided to build the National ICT Network 112, which would connect these services. The ICT systems of rescue services (including the Police systems) that are part of the ICT Emergency Notification System communicate via a unified interface. The basic functionalities of the Police Command Support System include:

- acceptance, registration, handling of notifications and events,
- management of activities, forces and resources of the Police,
- information management, including checks in the National Police Information System, publishing police messages,
- visualization of reports, actions and the possibility of using the police forces to handle the tasks performed,

- the use of mobile terminals for cooperation with the Command Support System and the National Police Information System,
- analytical system, reporting and statistics,
- as part of the emergency notification system, cooperation with the ICT system of the Emergency Notification Center and the ICT system of the State Fire Service.

3.3   Operational Information System in the Police

The Operational Information System is administered and maintained by an organizational unit within the structures of the General Police Headquarters, i.e. Bureau of Intelligence and Criminal Information (DPC. no 338, 2016). The bureau operates on the basis of an ICT security accreditation certificate, in the manner specified in the provisions on the protection of classified information (PCI, 2010). The Operational Information System processes information obtained by police officers, including personal data useful for the prevention, recognition, disclosure and detection of crimes, determining the methods of committing them, as well as detecting and detaining their perpetrators, in particular:

(1)   information on events, places, vehicles, documents and natural persons and entities other than natural persons;
(2)   information on telecommunications terminal equipment used to transmit information;
(3)   information on accounts at banks or other financial institutions and on banking activities.

The Operational Information System consists of three data sets: the Information Reporting System, the Central Information Base and the Archive Database. The databases contained in the Information Report System and the Central Information Base on Findings are intended to collect information used by the criminal and investigative service of the Police in connection with the performance of operational and exploratory activities, while the Information Report System is also intended to check and register places during the performance of operational, exploratory and investigative activities. The Archive Database is intended to collect information processed in the Information Report System and the Central Information Base on Findings, which have been deemed unsuitable for the performance of statutory Police tasks, in order to remove them.

3.4 Emergency Vehicle Driving Simulator in the Process of Training Polish Police Officers

Simulators, already widely used in aviation, are increasingly implemented in the process of training and psychophysical testing of vehicle drivers. Simulators of emergency vehicles are a perfect complement to a real vehicle training, providing unique opportunities:

- a safe training of vehicle driver behavior in dangerous situations;
- no possibility of creating a real threat to other road users;
- the ability to drive and conduct training in various terrain and weather conditions, regardless of external factors;
- repeatability of the exercises performed, ensuring the comparability of their results for the trainees.

One of the groups of people who should have special driving skills are the drivers of emergency vehicles used in the performance of official tasks, including by the Police, State Fire Service, the Military Freeware and other services.

A modern training tool called *Emergency vehicle driving simulator during typical actions and emergencies*, at the disposal of the Police Academy in Szczytno, is a simulation training system intended for training officers of various services in the field of emergency vehicle driving. The simulator is also used to train employees of companies dealing with the protection of persons and property as well as civilian drivers. Apart from improving the driving technique, the simulator enables, first of all, driving training in simulated conditions of various hazards (vehicle fire, explosion of a "trap car", tank fire and many others), as well as learning tactical team driving of persons moving in a column or those who conduct pursuit-blockades.

The simulator consists of the following modules:

- instructor and operator station;
- passenger car cabin located on a movable platform with six degrees of freedom, with an image presentation system with a field of view of 240° horizontally and 40° vertically;
- stationary cabin of a delivery vehicle with an image presentation system with a field of view of 200° horizontally and 40° vertically;
- three simplified driver stations, each equipped with three monitors, a steering wheel with pedals and a gear lever;
- communication system;
- computer system.

Such a modular structure allows for easy expansion of the system with additional stations. In the basic version, the simulator allows for training up to 5 drivers at the same time (two in fully equipped cabins and three in simplified positions).

This enhances the development of the ability to move in columns, convoys, cooperation in the implementation of tasks and communication between training participants. Thanks to these possibilities, cooperation between several drivers can be easily trained.

The training takes place in a three-dimensional, fully realistic virtual environment, which may include real or synthetic terrain, and all participants of the exercise work in a common virtual environment. The image generated by the computer system is seen by individual participants of the exercise according to their current location. The virtual environment includes buildings with complete infrastructure and reflects the most typical conditions encountered on roads. It represents characteristic elements of the landscape with vegetation, streets with their infrastructure and surrounding buildings, other road users (vehicles, pedestrians), as well as participants of public gatherings or sports events in the form of groups of people or crowds.

In addition, all other elements of the virtual environment are simulated, e.g.

- seasons (summer and winter conditions);
- time of day (day, night, dawn and dusk);
- weather conditions (rainfall, snowfall and haze of different intensity levels, wind speed and direction, etc.).

The software that constitutes the module of artificial intelligence also plays an extremely important role in the simulation. It is responsible for the interaction of individual simulated objects, in particular of guided vehicles and traffic. Algorithms of artificial intelligence are also used to simulate pedestrian traffic, including pedestrians moving in groups. The system allows not only for the creation of a significant number of event scenarios, but also for their adjustment to the individual level of knowledge and experience of the trainees. Furthermore, the instructor can modify the previously prepared scenario also during the exercise using the so-called dynamic changes of the operating situation.

A special module which is installed in the simulator is responsible for simulating the dynamics of vehicle movement and their collisions with surrounding objects, including pedestrians or animals. The instructor can choose from several types of vehicles most frequently used by different services, thus changing the dynamic characteristics of the simulated vehicle. While driving, vehicles may experience breakdowns introduced by the instructor or the system operator. The list of possible failures includes more than 30 vehicle failures, including failure of the ABS and ESP systems, the brake system, a drop in air pressure in the tires, play in the steering system, fire, etc.

The impression of realism of driving a car is intensified by a high level of fidelity to the environment, spatial sound background, the traffic system ensuring the appropriate sensation of movement, as well as the use of communication means used by the police and other services on a daily basis.

Each exercise on the simulator is preceded by the process of preparing its scenario and entering it into the system. It includes, inter alia, the determination of the location of activities and the initial course of events, time of the year and time of day as well as weather conditions. The degree of complexity of the activities may be each time adjusted to the level of training of people participating in a given exercise.

During the exercise, the instructor monitors its course and constantly adds additional elements of the tactical situation that directly affect the current circumstances during the exercise. The instructor can change, among others weather conditions, the degree of aggressiveness and the direction of movement of the participants of the assembly, as well as the staging of surprising events, such as all types of breakdowns, traffic accidents, initiating arson and fires, as well as terrorist activities, including the use of explosives.

As a result, it is possible to create a variety of decisive situations based on the same event scenario. Consequently, the process of analyzing a dynamically changing situation, risk assessment and making the right decisions is very similar to the real conditions. Each exercise is recorded, thanks to which it is possible to conduct a full analysis of its course and assess the effectiveness and efficiency of actions taken by individual participants of the exercise.

Apart from training functions, the simulator can also perform research functions, allowing to track the behavior and reactions of the drivers exercising in real time. It is possible thanks to the installation of a system that tracks the eye movement of the driver and a system for monitoring his basic vital parameters (e.g. blood pulse, sweating) with which the simulator is equipped. Thanks to this, it is possible to assess the predisposition of individual drivers to their profession and check how they react in dangerous situations related to their profession.

The simulator is characterized by an open architecture that allows for its extension by expanding the terrain database, introducing new objects, creating additional scenarios, as well as increasing the number of positions and types of vehicles.

3.5   Use of Wearable Cameras by Polish Police Officers

On the basis of the rights under Art. 15 s. 1 point 5a of the Act on the Police of 6 April, 1990, since 2019 throughout the country, police officers use wearable cameras to record official activities (PA, 1990). The fixed recordings are kept by the Police for 30 days from the date of recording. In cases where the recordings will be used for evidence purposes, this period will be extended by the period specified in separate regulations. The processing of data recorded with the use of wearable cameras meets the requirements set out in Art. 31 of the Act of 14 December, 2018 on the protection of personal data processed in connection with the prevention and combating of crime (PPDA, 2018).

The device transmits video and sound in the encryption system. From the moment of recording the event, the camera identifies a police officer and sets the time and date of intervention. The officer is not allowed to interfere with the recording. After the service, the police officer inserts the camera in the docking station. This archives the recorded data. The cameras are attached to uniforms with very strong magnets, which makes it practically impossible to lose them during the intervention. Before intervening, police officers inform that it will be recorded. Then they turn on the button that activates the camera.

When it comes to technical parameters, wearable cameras have three operating states, i.e. off, buffering mode, i.e. loop recording and continuous recording. They allow you to record in a quality of at least 1080p25 with the possibility of changing the resolution to 720p25 or 720p30. The internal memory allows for a minimum of 10 h of recording in 720p25 or 720p30 quality or a minimum of 8 h of recording in 1080p25 quality. Wearable cameras are adapted to recording in low light conditions, maintaining parameters comparable to the human eye. All recordings are encrypted, and the files recorded on the camera cannot be played back on a computer without dedicated software. Currently, the Polish police has almost 3,000 wearable cameras, which are used by officers of the prevention service during their daily service.

3.6 Directions for the Modernization of ICT Equipment in the Polish Police in 2021

The most important activities related to the modernization of the central ICT and ICT systems of the Polish police, planned for implementation in the years 2021–2025, include:

- construction of a covert communication system through the delivery and configuration of cryptographic devices;
- expansion of the ICT service monitoring system in international systems;
- expansion of the CybeArk PAM privileged account access system along with the purchase of administrative positions;
- expansion of WAF F5 infrastructure for two centers for a minimum period of 36 months;
- expansion of systems supporting the processing of fingerprints, including the Eurodac and Pruem interface. The system will be connected with the FBI (Federal Bureau of Investigation), and the cost of these activities will amount to over 50 million Polish zlotys;
- expansion of the National Police Information System and purchase of a license for the newly created Register of Offenses;
- building a cybersecurity system;
- modernization of police radio networks in 13 cities and urban agglomerations to the ETSI Tetra standard system.

In addition, at Polish police research centers, i.e. the Police Academy in Szczytno and the Central Forensic Laboratory of the Police in Warsaw, work is underway to create an AI system that will process data (including biometric data) and images obtained by, among others police drones, by carrying cameras by officers and stationary cameras installed in police vehicles, to identify people, vehicles, things.

The budget of the Polish police allocated annually to the so-called material expenditure (including the modernization of ICT equipment and development of ICT infrastructure) amounts to over PLN 2 billion per year.

4 ICT Systems Used by Polish Secret Services

4.1 Pegasus Platform in the Activities of the Central Anti-Corruption Bureau

The Central Anticorruption Bureau is a special service established to combat corruption in public and economic life, in particular in state and local government institutions, as well as to fight with activities detrimental to the economic interests of the state. It operates on the basis of the Act of 9 June, 2006 on the Central Anticorruption Bureau and the activities of the service are financed from the state budget (CABA, 2006). Currently, the Central Anticorruption Bureau employs over 1,300 police officers and civil employees who are high-class professionals in the fields related to the implementation of the tasks of the service. They are specialists in the area of operational work, investigative activities, analyses, control activities, direct protective measures, security of ICT networks, protection of classified information and personal data, legal service, logistics and finance, audit and internal security, international cooperation and security and staff training.

Within the limits of statutory tasks, officers of the Central Anticorruption Bureau perform:

- operational-exploratory activities aimed at preventing, recognizing and detecting crimes, and—if there is a justified suspicion that a crime has been committed—investigative activities aimed at prosecuting offenders;
- control activities aimed at revealing corruption cases in state institutions and local self-government and abuses of persons performing public functions, as well as activities detrimental to the economic interests of the state;
- operational, exploratory, analytical and informative activities in order to obtain and process information relevant to combating corruption in state institutions and local self-government and activities detrimental to the economic interests of the state.

The purchase and use of the Pegasus platform, created by the Israeli company NGO Group are kept in strict confidence. According to media reports, the Central Anti-Corruption Bureau acquired the Pegasus platform in 2018 for 33 million Polish zloty. It enables interception of data transmission without the operator's knowledge and it means that the platform owner is able to eavesdrop on phone calls and read messages sent via instant messaging. In addition, Pegasus has considerable ability to infect even modern smartphones. Installing the application sometimes requires clicking on a crafted link, other times it does not require any interaction on the part of the user. Many civic organizations are worried about such large application possibilities. The tool used to prosecute criminals and terrorists may also be used in an unauthorized way to surveillance politicians and journalists.

4.2 ICT- and AI-Enhanced Tools and Solutions Used by the Internal Security Agency

The Internal Security Agency is a special service whose task is to protect the state against planned and organized activities that may pose a threat to Poland's independence or constitutional order, disrupt the functioning of state structures or jeopardize the basic interests of the country. The Internal Security Agency fulfills its duties, among others by obtaining, analyzing and processing information about dangers. Completed studies are submitted to the relevant constitutional organs. In carrying out its tasks, the Internal Security Agency uses operational and process powers. The catalog of the basic tasks of the Internal Security Agency includes: combating terrorism, counter-espionage, counteracting the proliferation of weapons of mass destruction, fighting economic crimes, fighting organized crime, fighting corruption, protecting classified information, and preparing analyzes and information about threats to security in internal. Currently, the Agency employs around 5,000 people.

The Internal Security Agency has a statutory obligation to identify terrorist threats and prevent acts of terror. The tasks of the Agency include acquiring and analyzing information, assessing the sources and scale of the phenomenon, selecting groups of potential bombers, identifying their plans and logistics. In the fight against terrorism, the Agency closely cooperates with other services and state institutions as well as international organizations. Effective coordination of activities undertaken by units responsible for anti-terrorist protection in Poland is ensured by the Anti-Terrorist Center located within the structures of the Agency. The Anti-Terrorist Center of the Internal Security Agency is a coordination and analytical unit in the field of counteracting and combating terrorism. The center operates 24 h a day, 7 days a week. The service is performed by officers of the Internal Security Agency and seconded officers, soldiers and employees, among others Police, Border Guard, State Protection Service, Foreign Intelligence Agency, Military Intelligence Service, Military Counterintelligence Service and Customs Service. They carry out tasks within the competence of the institution they represent. Furthermore, other entities participating in the anti-terrorist protection system of Poland, such as the Government Centre for Security, the Ministry of Foreign Affairs, the State Fire Service, the General Inspector of Financial Information, the General Staff of the Polish Army, Military Police, etc. actively cooperate with the Anti-Terrorist Center.

The essence of the functioning of the Anti-Terrorist Center of the Internal Security Agency is the coordination of the information exchange process between the participants of the anti-terrorist protection system, enabling the implementation of joint response procedures in the event of the occurrence of one of the four categories of defined threat:

- a terrorist event outside Poland affecting the security of the country and its citizens;
- a terrorist event in Poland affecting the security of the country and its citizens;
- obtain information about potential threats that may occur in Poland and abroad;
- obtaining information on money laundering or transfers of funds that may indicate financing of terrorist activities.

The Internal Security Agency uses to the greatest extent all databases and ICT systems that collect data of people, things, vehicles, and facilities which are at the disposal of state administration bodies in Poland. The agency uses, among others systems: the Universal Electronic System of Population Register, Central Register of Vehicles and Drivers, National Police Information System, Archival Resource Record System, Central Statistical Office database, Electronic Services Platform of the Social Insurance Institution. In total, the Internal Security Agency has officially access to over 40 national databases. This service also has access to all state secrets and solutions in the field of ICT, applied computer science and AI currently available to Poland. Currently, the Internal Security Agency participates in the work of an international scientific and research consortium, which, from the funds of the European Union budget, is building a system of biometric identification of people and vehicles for the needs of, inter alia, protection of the Union's external borders.

5   Discussion

The main issue in this chapter is the impact of actions taken by police formations in Poland with the use of the latest ICT-enhanced tools and solutions, on the respect for fundamental civil rights and freedoms included in the most important legal acts, i.e. the Constitution of the Republic of Poland and the European Convention on Human Rights. The discussion in this chapter suggests that the key problem/question to be addressed is this one: Is the consent of citizens to limit their rights and freedoms not a sine qua non condition for maintaining a sense of security in the face of the existing threats to public safety in Poland? According to public opinion polls in Poland, the Police are a uniformed force that is very well assessed by Poles. Over 75% of respondents evaluate the work of police officers positively, and every sixth (16%) is dissatisfied with it (CBOS, 2019). Every day, Polish police officers make about 17.5 thousand interventions. As part of the investigation work, the Polish police conduct over 760,000 proceedings (preparatory proceedings, investigations) in cases of crimes every year. In order to detect the perpetrators of these crimes, modern ICT solutions are also implemented, and they allow for the identification or infiltration of criminal groups. Activities with the use of ICT-enhanced systems also carried out by other Polish police services are characterized by a high degree of use of this type of ICT-enhanced means.

The activities of police formations with the use of ICT may result in violations of civil rights and freedoms in connection with interference with the right to respect for private or family life, home or correspondence within the meaning of art.8 of the European Convention on Human Rights and art. 47 (legal protection of private life) and 49 (protection of the confidentiality of communication) of the Polish Constitution. It should be noted that Article 8 of the European Convention on Human Rights allows cases of interference with the right to respect for private life, in point 2, when it is provided for by another legal act of the rank of an Act. Similarly, the Constitution of the Republic of Poland states that the limitation of the rights included in art. 47 and 49 can occur exclusively in a manner specified by the Law.

Violation of the right to respect for private life may take place, for example, when a telephone wiretap is installed, without entering the apartment being wiretapped (e.g. drilling through the wall, inserting the wiretap in the ventilation system). Control of "correspondence" (including electronic correspondence) is of course closely related to both "private life" and "family life", and the scope of its application is so broad that it covers most means of communication. Therefore, several techniques used to prevent or prosecute crime, such as intercepting messages (MvUK, 1984), using electronic eavesdropping devices (KvUK, 2000) monitoring e-mails (CvUK, 2007), and using GPS trackers (UvG, 2010), incorporate interference with the rights laid down in art. 8, even if the data obtained in this way will never be used later (KvS, 1998). However, when it comes to gathering intelligence through observation in public places, art. 8 only applies when a state action goes beyond mere observation and includes active monitoring of individuals (PvUK, 2003). Interference with the rights defined in art. 8 must meet a number of conditions to be considered justified under the European Convention on Human Rights. These conditions are:

- interference must be justified by pursuing a legitimate aim;
- provided for by law;
- necessary in a democratic society.

If the state cannot meet any of these conditions, it will be considered as a breach of the guarantee. Such conduct of the police services must be an action necessary in a democratic society for the sake of state security, public security or the economic well-being of the country, the protection of order and crime prevention, the protection of health and morals or the protection of the rights as well as freedoms of persons.

It is significant for police action that appropriate measures are in force in order to ensure compliance with both national law and the European Convention on Human Rights. For example, when the national law-permissible enforcement of police powers covers a wide range of cases, enforcement officers should ensure that they are used only where there is a clear need and for the purpose for which they are intended. Such proceedings may reduce the likelihood of their actions being effectively challenged before national courts or the Strasbourg Court.

ICT measures most often implemented by Polish police units in the case of activities interfering with the right to respect for private life include telephone wiretaps, surveillance with the use of optical and electronic devices, control of electronic correspondence, including accounts on social networks, control of bank accounts, collection and processing of data using modern spy software enabling, for instance, access to all phone content, as well as to the loudspeaker and the camera. According to the statistical data made available. to the public by the Polish law enforcement authorities, only in 0.002% of the cases of actions taken (in 2017) with the use of ICT measures, the courts raised their objections as to the fulfillment of formal conditions for the implementation of such measures in investigative activities. This proves well-thought-out and reliable activities with the use of ICT. It should be emphasized that in virtually every case conducted by police formations in Poland, information contained in ICT databases is applied, and their collection, storage and processing is carried out strictly in accordance with the guidelines and regulations deriving from many legal acts, e.g. from the Act on Police, Act on the Protection of Classified Information, Act on Personal Data Protection or Regulation (EU) 2016/679 of the European Parliament and of the Council of 27 April 2016 on the protection of natural persons with regard to the processing of personal data and on the free movement of such data, and repealing Directive 95/46 / EC (General Data Protection Regulation).

In order to fully seal the system of implementing ICT measures in activities carried out by police formations, which in essence violate civil rights and freedoms, apart from the multi-faceted system of external (institutional) control and supervision, a system should also be developed (a legislative solution correlated with the technical and organizational procedure), which, on the one hand, would protect civil rights and freedoms and, on the other hand, ensure the security of citizens (Dycus et al., 2016, p. 32; Gearty, 2005, p. 27; Head, 2002, p. 671; Macken, 2011, p. 91).

6 Conclusion

The electronic means of collecting, processing, and transmitting information are synonymous with today's civilization. We are all witnesses and participants in the creation of a new category of society in the interactive world, the computerized society, whose commodity is information. This society of the age of computers and computer networks is known as the information society (cf. Visvizi et al., 2021).

In the local, regional, or global dimension, information is the most important component that creates the state of security. Therefore, all entities and institutions strive to obtain and process as much information as possible, which will allow for a quick and adequate response to emerging threats. In this regard, it is very important to have effective ICT tools which, having the features of AI, will allow for the implementation of proactive activities, thanks to which the risk of threats both in terms of internal security (e.g. increased crime) and external security (e.g. protection of economic interests of the state) is minimized. Taking care of its security, Poland invests more and more funds each year in the development of its ICT infrastructure, including AI solutions, thanks to which the level of security of the country increases, not only on a local but also on a regional level. In addition to expanding its own ICT capabilities, strengthening the potential of public safety services, and investing in defense potential, Poland's international cooperation, including membership in the

European Union and NATO, is also a leading element in increasing the country's security.

References

Alleweldt, R., & Fickenscher, G. (Eds.). (2018). *The police and international human rights law.* Springer.

Bowling, B., & Sheptycki, J. (2012). *Global policing.* Sage.

CABA. (2006). Act of 9 June, 2006 on the Central Anticorruption Bureau, Journal of Laws of the Republic of Poland 2006, No. 104, item 708.

CPHR. (1950). Convention for the Protection of Human Rights and Fundamental Freedoms drawn up in Rome on 4 November, 1950, later amended by Protocols No. 3, 5 and 8 and supplemented by Protocol No. 2, Journal of Laws of 1993, No. 61, item 284.

CvUK. (2007). Judgment of 3 April 2007 in Copland versus United Kingdom, paragraphs (pp. 41–42).

DPC. (2013). Decision no. 125 of the Police Commander in Chief of 5 April, 2013 on the operation of the National Police Information System, § 13, section 1.

DPC. (2016). Decision No. 338 of the Police Commander in Chief on the Operational Information System of 12 October, 2016.

Dycus, S., Bank, W. C., Raven-Hansen, P., Vladeck, S. I. (2016). Counterterrorism Law. Wolters Kluwer, Alphen aan den Rijn.

GDPR. (2016). Regulation (EU) 2016/679 of the European Parliament and of the Council of 27 April 2016 on the protection of natural persons with regard to the processing of personal data and on the free movement of such data, and repealing Directive 95/46/EC (General Data Protection Regulation).

Gearty, C. (2005). 11 September 2001, Counter-terrorism, and the Human Rights Act. *Journal Law and Society, 32*(1), 18–33.

GP. (2020). http://gazeta.policja.pl/997/inne/prawo/87063,KSIP-nowe-zasady-przetwarzania-inf ormacji-Nr-98-052013.html. Accessed on 12 Aug 2020.

Grieve, J., Harfield, C., & MacVean, A. (2007). *Policing.* Sage.

Guidelines on human rights education. For law enforcement officials, Published by the OSCE Office for Democratic Institutions and Human Rights (ODIHR), Warsaw, 2012.

Head, M. (2002). Counter-terrorism' laws: a threat to political freedom, civil liberties and constitutional rights [Critique and Comment.]. *Melbourne University Law Review, 26*(3), 666–689.

https://www.cbos.pl/SPISKOM.POL/2019/K_044_19.PDF. Accessed on 10 Nov 2020.

ICCPR. (1966). International covenant on civil and political rights opened for signature in New York on 16 December, 1966, Journal of Laws of UN 1977, No. 38, item 167.

KvS. (1998). Judgment of 25 March 1998 in Kopp versus Switzerland, paragraphs pp. 51–53.

KvUK. (2000). Judgment of 12 May 2000 in Khan versus The United Kingdom, paragraph 25 (negotiations with wiretapping of conversations).

Langford, M. (2007). The right to social security and implications for law, policy and practice. In: E. H. Riedel (Ed.), *Social security as a human right. drafting a general comment on Article 9 ICESCR—some challenges* (pp. 29–53). Springer.

Macken, C. (2011). *Counter-terrorism and the detention of suspected terrorists: Preventive detention and international human rights law.* Routledge.

MvUK. (1984). Judgment of 2 August 1984 in Malone versus The United Kingdom, paragraph 64 (telephone wiretapping).

PA. (1990). Act of 6 April, 1990 on the Police, Journal of Laws of 2019, item 161.

Pawłowicz, M. (2014). Organizacyjno-prawne kierunki zmian zarządzania projektami teleinfor-
matycznymi w Policji [Organizational and legal directions of changes in the management of ICT
projects in the Police]. Dissertation, Police Academy in Szczytno.

Pawłowicz, M. (2016). Nowa jakość projektów teleinformatycznych a bezpieczeństwo organizacji
policji [New quality of ICT projects and the security of police organization] Publishing House
of the Police Academy, Szczytno.

PCI. (2010). Act of 5 August 2010 on the protection of classified information, Journal of Laws of
the Republic of Poland 2010, No. 182, item 1228.

PPDA. (2018). Act of 14 December, 2018 on the protection of personal data, Journal of Laws of
the Republic of Poland 2019, item 125.

Project *Legal and criminological aspects of the implementation and use of technologies for
the protection of internal security*, carried out in 2011–2012 by a consortium: Polska Plat-
forma Bezpieczeństwa Wewnętrznego Sp. z o.o. [The Polish Platform of Internal Security Ltd.]
(the leader) and the University of Bialystok. (Project Number 0R00037, Agreement Number
0037/R/T00/2009/2007).

Pływaczewski, E. W. (2017). Bezpieczeństwo osobiste. Prawa człowieka. Zrównoważony rozwój
[Security of citizens. Human Rights. Sustainable development], Temida 2 Publishing, Białystok.

PvUK. (2003). Judgment of 28 January 2003 in Peck versus The United Kingdom, paragraphs 57
to 63.

Regulation no. 63 of the Police Commander in Chief of 7 October, 2019 on the detailed rules for
the use of unmanned aerial vehicles in the Police (Official Journal of the Police Headquarters of
2019, item 106).

RP. (1997). Constitution of the Republic of Poland of 2 April, 1997, adopted by the National
Assembly on 2 April, 1997, adopted by the Nation in its constitutional referendum on 25 May,
1997, signed by the President of the Republic of Poland in on 16 July, 1997.

RPO. (2020a). https://www.rpo.gov.pl/pl/content/etpc-zbada-uprawnienia-polskich-sluzb-specja
lnych-opinia-rpo. Accessed on 6 Aug 2020.

RPO. (2020b). https://www.rpo.gov.pl/pl/kategoria-konstytucyjna/art-31-wolnosc-i-dopuszczalne-
ograniczenia. Accessed on 7 Aug 2020.

UvG. (2010). Judgment of 2 September 2010 in Uzun versus Germany, paragraphs 49 to 53
(observation with GPS tracking devices).

Visvizi, A. (2015). Safety, risk, governance and the Eurozone crisis: Rethinking the conceptual
merits of global safety governance. In P. Kłosińska-Dąbrowska (Ed.), *Essays on global safety
governance: Challenges and solutions* (pp. 21–39). University of Warsaw, ASPRA-JR.

Visvizi, A. (2016) The conceptual framework. In A. Visvizi, T. Stępniewski (Ed.), *Poland, the Czech
Republic and NATO in fragile security contexts* (pp. 13–15). IESW Reports.

Visvizi, A., Lytras, M. (Eds.) (2019). Politics and Technology in the Post-Truth Era, Bingley, UK:
Emerald Publishing, ISBN: 9781787569843, https://books.emeraldinsight.com/page/detail/Pol
itics-and-Technology-in-the-PostTruth-Era/?K=9781787569843.

Visvizi, A., Lytras, M. D., Aljohani, N. R. (Eds). (2021). Research and Innovation Forum 2020:
Disruptive technologies in times of change, Cham: Springer. https://doi.org/10.1007/978-3-030-
62066-0, https://www.springer.com/gp/book/9783030620653.

Jacek Dworzecki, Ph.D., Professor Academy Police Force in Bratislava, and Military University of Land Forces in Wroclaw, Poland. Former police officer. Expertise: security and combating
crime. Head of the Polish and Slovak research teams in the frame of international scientific
projects, e.g., TAKEDOWN, H2020-FCT-2015, No: 700688. ORCID ID: 0000-0002-9357-5713

Izabela Nowicka, Ph.D., Associate Professor Military University of Land Forces in Wroclaw,
Poland. Her areas of expertise include criminal law, military law and legal acts in the field of
public safety. ORCID ID: 0000-0001-8974-0803

The United States of America's Embrace of Artificial Intelligence for Defense Purposes

Mikołaj Kugler

Abstract Artificial intelligence (AI) is used across an increasingly wide range of sectors. Given its transformative potential, AI carries profound implications in a national security context, especially as regards increasing countries' military advantage. The key leaders in AI development, and at the same time adversaries and competitors for global supremacy and leadership, such as China, Russia and the United States (US), have recognized this potential inherent in AI and are developing initiatives to adopt AI in pursuit of their national security goals. Thus, AI has become yet another area of great power rivalry. Given the above, this chapter, first, explores the ways in which AI can impact national security in a military context. Then, it presents how China and Russia endeavor to boost their military competitiveness with AI-enabled capabilities. Finally, it juxtaposes the approach adopted by the US to maintain its strategic position in the world and remain the key security provider for its citizens and the US allies with approaches adopted by the US' rivals.

Keywords Artificial intelligence (AI) · The United States of America · National security · Military advantage · National power · Great power rivalry

1 Introduction

AI is a rapidly expanding field of technology, with applications across the full spectrum of human activity. As such, it also carries far-reaching implications for national security. One example of its high-profile application is the United States (US) military's Project Maven in Iraq and Syria (Pellerin, 2017; Seligman, 2018; West & Allen, 2018). Other military functions of AI discussed in the literature include intelligence, logistics, cyberspace operations, information operations, command and control, semiautonomous and autonomous vehicles, and lethal autonomous weapon systems (cf. CRS, 2020, p. 9–15; Sheppard et al., 2018, p. 27–30).

M. Kugler (✉)
General Tadeusz Kościuszko Military University of Land Forces, Wrocław, Poland
e-mail: mikolaj.kugler@awl.edu.pl

© The Author(s), under exclusive license to Springer Nature Switzerland AG 2021
A. Visvizi and M. Bodziany (eds.), *Artificial Intelligence and Its Contexts*, Advanced Sciences and Technologies for Security Applications,
https://doi.org/10.1007/978-3-030-88972-2_12

The potential that AI carries will influence the security environment and, thus, will have a profound impact on the rivalry for global leadership. In 2017, China released its strategy in which it declared its ambition to lead the world in AI by 2030 (China State Council, 2017). Similarly, Russia's President Vladimir Putin stated that the nation leading in the field of AI will become the ruler of the world (Gigova, 2017). Likewise, the 2018 US National Defense Strategy has placed AI among the very technologies that ensure that the US will be able to fight and win the wars of the future (DoD, 2018). What once used to be a technical and academic issue has now become the focus of geopolitical competition (Buchanan, 2020). Claims of an AI arms race appear fully justified.

China's and Russia's actions towards the US are described as adversarial, with both countries pushing the boundaries and attempting to reassert their influence regionally and globally (WH, 2017, p. 27) as well as exerting pressure on the US to compete for innovative military AI applications (CRS, 2020). With both countries undermining the US primacy, its competitive edge and posing a threat to the liberal values and norms the US and its allies live by, it seems imperative that the US leadership in the field should top American national security agenda (Schmidt & Work, 2019), with requisite measures taken to accomplish this goal (cf. Sheppard et al., 2018, p. 4). However, there are concerns that whereas China and Russia are making progress in the field, the US may be lagging behind (Auslin, 2018). Moreover, despite some steps being taken by the US government, there is no comprehensive strategy that would set forth the US policy (New, 2018, p. 1; Sherman, 2019; Groth et al., 2019, p. 7). Considering the above, the objective of this chapter is to examine how the US embraces AI for defense purposes so that it maintains its strategic position and is still capable of advancing security, peace, and stability. The chapter is structured as follows: first, it elaborates on AI's potential impact on national security; next, it presents how China and Russia encompass AI-enabled defense capabilities; then, it discusses the US approach to AI military integration. Conclusions follow.

2 AI and National Security

AI is here to stay. It has been changing the way we live our lives for quite some time now and it continues to develop rapidly, finding its application in more and more everyday technologies as well as across a number of sectors. In addition, under the concept of AI for the social good, it is also employed to deal with major societal challenges (CRS, 2017).

As AI is a game changer across a wide range of businesses and industries, it is believed that it is bound to significantly impact national security in a variety of ways, not only transforming the character of the future battlefield and the pace of threats, but also affecting the balance of power in both the global economy and military competition (Sheppard et al., 2018, p. 1). Technology has always played a key role in achieving supremacy in power over others. It has always been a key factor that could tip the scales in one side's favor. Security has always been accelerated by technology. And so have threats (cf. WH, 2017, p. 26).

There are numerous studies and scenarios prognosticating AI's impact on national security (Allen & Chan, 2017; CRS, 2020; DIB, 2019; NSCAI, 2019; Nurkin & Rodriguez, 2019). What they all appear to acknowledge is that the capabilities that AI can already provide carry significant potential for national security should be viewed beyond their impact on a particular military task and will have strategic implications. It is even claimed that AI's potential is as transformative and impactful for national security as that of nuclear weapons, aircraft, computers, and biotech, predicting that AI will affect national security by causing changes in the following three areas (Allen & Chan, 2017, p. 1–3):

1. military superiority, as not only will progress in AI enable new capabilities, it will also make the existing ones more affordable to a wider audience, such as a weak state or a non-state actor acquiring a long-range precision strike capability owing to a purchase of a commercially available AI-enabled long-range unmanned vehicle;

2. information superiority, as AI will considerably boost the collection and analysis of data, improving the quality of information available to decision makers, as well as their creation, thus facilitating deception and the distortion of truth, which may lead to undermining trust in many public institutions;

3. economic superiority, as AI might spark off a new industrial revolution, for instance as a result of dwindling demand for labor.

Given AI's capabilities, it can be assumed that its applications can benefit national security in the following ways (Nurkin & Rodriguez, 2019, p. 24):

1. by enabling humans, i.e. enhancing operators, intelligence officers and strategic decisions makers and the like dealing with exponentially growing massive troves of data and information in the performance of their tasks, particularly humdrum and long-duration ones, such as intelligence collection and analysis, this increasing their productivity and endurance;

2. by removing humans, i.e. replacing them, for instance with unmanned systems, in the execution of dirty, mundane or hazardous tasks, such as cleaning contaminated environments or handling explosives;

3. by exceeding humans, i.e. facilitating the development of new capabilities characterized by almost full autonomy and minimal human involvement, incredibly fast reaction times and unparallel processing power, resulting in greater situational awareness, accelerated pace of combat or a force-multiplying effect, such as a swarm of drones being potentially able to overpower a highly-advanced weapon system.

AI-enabled operational capabilities that appear to be contributing in general to the achievement of national security and military objectives, as they are being developed and fielded by a variety of actors, include the following (Nurkin & Rodriguez, 2019, p. 23):

1. Enhancing Processing, Cognition, and Decision-Making: coping with big data and enhancing processing and cognition;
2. Simulation and Training: simulating complex environments and behaviors, evaluation of training outputs, AI as a tutor—improving training efficiency;
3. Autonomous platforms and Systems: autonomous platforms, swarms, teaming mother ships and loyal wingmen, lethal autonomous weapons systems;
4. Human Performance Enhancement: human–machine intelligence fusion, pilot support, exoskeletons and AI;
5. Logistics and Maintenance: predictive maintenance to reduce costs and extend the lifetime of platforms;
6. Sensors, Communications, and Electronic Warfare (EW): cognitive sensing, radios and radars, cognitive EW;
7. Competition in the information Domain: cyberattack and defense, disinformation campaigns and influence operations;
8. Security and Surveillance: border and event security, targeted surveillance, social credit score support.

The above list is not finite and is bound to expand along with progress in AI research, the further development of technology and the wider adoption and incorporation of AI by individual actors.

On the whole, the employment of AI in the national security context is to facilitate and embrace the vast feeds of data and information available from different sources and geographic locations for a range of mission- or task-specific solutions, gradually eliminating the human component. It will allow national security organizations to understand and execute their missions better and faster. The fusion of AI with military systems will increase the accuracy and speed of perception, comprehension, decision-making, and operation beyond the capability of human cognition alone. The impact is expected to be so profound that some are convinced that AI will lead to the inception of so-called "algorithmic warfare", in which algorithms will fight against algorithms, with the speed and accuracy of knowledge and action carrying more weight than standard factors such as the number of troops or firepower. Those with unparalleled data, compute power and algorithms will gain unprecedented battlefield advantage (NSCAI, 2019, p. 10).

The successful adoption of AI and its integration into the military is not unproblematic, however, and does pose certain challenges. Most of them stem from the fact that unlike in the past, when it was the government-directed defense-related programs that inspired new technologies subsequently transferring to the civilian sector, currently commercial companies are essentially at the forefront of AI development. This is largely due to government agencies seriously deficient in adequate expertise and resources, predominantly AI talent. As a result, the military is left with the necessity to eventually acquire and adapt commercially-developed tools for its defense applications. And it is fraught with a range of additional impediments related to complicated and lengthy government acquisition and procurement procedures, reluctance to partner with and collaborate with the military due to, for instance, ethical concerns over the use of AI in surveillance or weapon systems, issues with intellectual property and data rights (US Government Accountability Office, 2017), or simply different work culture and mutual distrust (CRS, 2020, p. 19).

When it comes to the actual adoption of commercially-developed AI technology, it should be borne in mind that certain applications will require only minor adjustments before they can serve their national security purpose, whereas others will necessitate quite profound modifications. As for the former, take predictive logistics, in the case of which it may suffice to only provide parameters for a particular piece of hardware. As for the latter, the more extensive customization will be required due to the differences between the environment for which the technology has initially been developed and the one for which it should be suited. This will be particularly true for autonomous or semiautonomous vehicles that have been or will be developed for operation in data-rich environments, e.g. with GPS positions or up-to-the-minute traffic information, while their military equivalents will need to be able to operate and navigate in rough, poorly-structured conditions, off-road, with incomplete terrain mapping or no GPS signal due to jamming (CRS, 2020, p. 16).

Another issue concerns standards of safety and performance, which in the case of the military are invariably high, but not necessarily so for the civilian ones. A failure rate regarded acceptable for a civilian deployment of AI technology may be found unacceptable in a military context due to strict requirements imposed on military systems. A particular AI-enabled solution can be adopted by the government only after high levels of trust and reliability have been ensured and any issues of operational control have been worked out (Sheppard et al., 2018, p. 27).

Last but not least, there also concerns regarding the safe operation of the technology itself, as AI algorithms may be unpredictable or susceptible to manipulation, bias and theft (CRS, 2020, p. 8), by being a potential target for any adversary, a state and non-state one alike, or by being trained on corrupt data. Algorithms based on distorted or biased data can yield unexpected or undesirable results (Layton, 2018, p. 13), thus generating various challenges that may be difficult to detect at tactical, operational, or strategic levels. Although AI has already surpassed humans in different contexts, the mistakes made are the ones that a person would never make, hence being hard to predict, prevent or mitigate (Kania, 2017, p. 44). This could be particularly consequential in a military context if such biases were incorporated into autonomous systems with lethal effects (CRS, 2020, p. 31), and if such systems were deployed at scale (Scharre, 2016, p. 23).

To sum up, AI provides numerous opportunities as well as challenges for its application within a national security context, and as such it demonstrates enormous transformative potential. Given its inherent dual-use nature, what impact AI will ultimately have on national security depends on the relationship between the government and the commercial AI community and the extent to which these two can work together, overcoming all obstacles, to maximize the technology's strengths and minimize its vulnerabilities.

3   China and Russia: Strategic Competition for the US

Given the above, it can be concluded that AI demonstrates considerable potential to affect the balance of power. This fact has already been acknowledged by China, Russia and the United States. On July 20, 2017 China's State Council released "A Next Generation Artificial Intelligence Development Plan", in accordance with which China should take the lead in AI by 2020 (China State Council, 2017). Shortly afterwards, Russia's President Vladimir Putin announced that his country is determined to pursue AI technologies, as this is a key to a global leadership position (Gigova, 2017). By the same token, the 2018 US National Defense Strategy included AI among key technologies, allowing America to fight and win the wars of the future (DoD, 2018, p. 3).

Regarding China, there is no denying that it is America's chief competitor in the international AI market. The abovementioned 2017 Chinese strategy describes AI as a "strategic technology" that has become a "new focus of international competition" (China State Council, 2017, p. 2). In pursuit of the strategic objectives established in the Plan, China is said to be planning to develop its AI's core industry of over 150 bn RMB (approx. $21 bn) by 2020 to 400 bn RMB (approx. $58 bn) by 2025 and 10 trillion RMB (approx. $1.5 trillion) by 2030 (Sheppard et al., 2018, p. 50). This appears to be corresponding to China's overall strategy to achieve global leadership in research and development. Interestingly, China's R&D funding increased 30 times from 1991 to 2015 and it is projected to overshadow the US in this regard within 10 years (NSCAI, 2019, p. 17).

China is conducting research, development and testing for a variety of AI-enabled military and security applications that will be critical to the future of conflict, including, but not limited to: intelligent and autonomous unmanned systems, such as swarm intelligence; AI-enabled data fusion, information processing, and intelligence analysis; applications in simulation, war-gaming, and training; the use of AI for defense, offense, and command in information warfare; and intelligent support to command decision-making (Kania, 2017, p. 21).

There have been reports depicting China's successful trails with the different types of air, land, sea, and undersea autonomous vehicles. For instance, in June 2017, China Electronics and Technology Group Corporation (CETC) demonstrated its advances in swarm intelligence by carrying out a successful test of 119 fixed-wing UAVs (Kania, 2017, p. 23). Another example is a 2018 test of a fleet of fifty-six unmanned vessels that, if equipped with weapons, could be used to attack enemy during sea battles (Barnes, 2018).

It is still an open question whether China will win its AI competition with the US. What might be seen as China's advantage is its unified, whole-of-government effort to develop AI. The Chinese government as well as the military, academic research laboratories, financial institutions, and corporations are aligned and work closely towards the common goal. As a result, the Chinese government can have primacy in setting AI development priorities and principles (Sheppard et al., 2018, p. 48–49), and can exercise central direction and control. Certainly, this can drive collaboration between the military and the civilian sector fast forward, thus attaining the AI development strategic objectives much faster.

On the other hand, what is recognized as a strength by some is thought to be a weakness by others. Despite the central management of China's AI ecosystem, it has been pointed out that its funding management is inefficient: corrupt, favoring some research institutions with government funding over others or even overinvesting beyond market demand. In addition, China faces the same problem as the US: a shortage of qualified personnel with sufficient experience in the field. Moreover, China also loses academically in AI to the US in terms of the number of AI programs run at Chinese universities and the quality of AI research and academic publications (CRS, 2020).

To judge by its President's words (Gigova, 2017), Russia is also determined to adopt AI, including for military purposes, although at the moment it appears to be behind both the US and China (Markotkin & Chernenko, 2020). In order to catch up with the competitors, in 2019 Russia released the "National Strategy for the Development of Artificial Intelligence Through 2030", detailing the way in which it is planning to enhance Russian AI expertise as well as educational programs, datasets, infrastructure and legal regulatory system. Interestingly, it does not make a direct mention of AI development for national security or defense purposes (Office of the President of the Russian Federation, 2019). This comes on top of Russian effort to modernize its defense forces, including the 30% robotization of its military equipment by 2025 (Simonite, 2017).

Russia's measures taken to close the gap with the United States and China include the establishment of a variety of organizations working on military AI development. In March 2018, a 10-point AI agenda was issued. Among other initiatives it propounds the establishment of an AI and Big Data consortium, a Fund for Analytical Algorithms and Programs, a state-funded AI training and education program, a dedicated AI lab, and a National Center for Artificial Intelligence (Bendett, 2018). What is more, Russia created the Foundation for Advanced Studies: a defense research organization dedicated to autonomy and robotics, and launched an annually-held conference on "Robotization of the Armed Forces of the Russian Federation (Bendett, 2017b). Moreover, in 2018 the ERA Military Innovative Technopolis was designated by the Russian Ministry of Defense its main AI research, development, test and evaluation (RDT&E) hub. In July 2020 it started accepting applications for science research competitions on artificial intelligence (CNA, 2020, p. 11).

Russia's research and development focus, apart from a variety of AI applications, has also been on semiautonomous and autonomous vehicles. In 2017 the chairman of the Federation Council's Defense and Security Committee stated that owing to AI it would be possible to replace a soldier and a pilot, further predicting equipping vehicles with AI (Bendett, 2017a). At this point it should be remembered that Russia has already conducted a successful test of an uninhabited ground vehicle that supposedly outdid existing [inhabited] combat vehicles. There are plans to deploy the system in combat, intelligence gathering, or logistics roles in the future (Davies, 2017). The Russian military is also developing and possibly already fielding advanced landmines that utilize some AI-enabled capabilities. It is said that the POM-3 (ПОМ-3) "Medallion" landmine allegedly had an ability to distinguish between various targets, for instance between a civilian and a soldier (CNA, 2020, p. 22). Like its competitors, Russia also has plans to deploy AI-enabled uninhabited vehicles. It is researching swarming capabilities and exploring other innovative uses of AI, for instance for electronic warfare. It should also be remembered that Russia has already employed AI technologies for propaganda and surveillance and is said to have directed information operations against the United States and its allies (CRS, 2020, p. 25).

The US-Russia competition is different from the one with China. Russia appears to pose a lesser challenge, nevertheless still a persistent one, mainly due to its AI-enabled cyber and information operations capabilities. If skillfully employed, they can serve as a force multiplier and can give Russia a competitive edge over its rival in an asymmetric or hybrid struggle.

Again, despite its major efforts and managing to make its mark in cyber and information operations, it also remains to be seen whether Russia will actually succeed in achieving its plans. Some experts are somewhat skeptical, considering fluctuations in Russian military spending. Moreover, they point out that Russia lacks firm academic base on which to build its progress in AI. It has not created AI applications of the quality comparable to those of the US and China either (CRS, 2020, p. 25). Success in AI-enabled military applications will require Russia's military to leverage its small but growing domestic AI industry. Russia's current (unclassified) investment levels in AI are significantly behind the United States and China, at approximately 700 million rubles ($12.5 million) (Polyakova, 2018).

4   US Policy Approaches to AI

The opportunities for defense purposes offered by AI-enabled technologies and the posture of the American international competitors, which might potentially lead to the erosion of US military advantage or undermining global stability and nuclear deterrence as such, have made it imperative for the US to eventually embrace AI and provide a strategic framework for its adoption that would leverage America's unique strengths.

It can be argued that AI will be central to American strength. AI-enabled technologies, which are at the center of power competition, are conducive to contesting the US primacy. According to one report published by the US-China Economic and Security Review Commission, advanced weapons systems enabled by AI are a "game-changer" and a "game-leveler" (Nurkin et al., 2018, p. 15). AI-enabled capabilities, more and more frequently employed on the battlefield, will eventually change the nature of conflict, allowing for tipping the scales in America's competitors' favor (Kallenborn, 2019).

The significance of US leadership in AI for the defense of the United States and the maintenance of the international order, and a concern that the US might be outpaced by its competitors have been asserted by US Senator Ted Cruz at a Senate hearing, who stated that allowing countries such as China and Russia to gain control in AI development might in the long run pose a real threat to US national security (US Senate, 2016, p. 2).

In general, it has found that (1) global leadership in AI technology is a national security priority, as it is vital to the future of American economy, society, and security, and (2) the adoption of AI for defense and security is a strategic necessity, as the U.S armed forces must have access to the most advanced AI technologies to protect the American people, allies and interests (NSCAI, 2019, p. 15).

The above has been reflected in the US strategy documents. The 2017 National Security Strategy of the United States of America listed AI among the technologies critical to economic growth and security, allowing the US to maintain its competitive edge (WH, 2017, p. 20). This idea has been sightly elaborated on in the follow-on document: The National Defense Strategy of the United States of America issued by the Department of Defense in 2018 (DoD, 2018). Without offering much guidance, the Strategy recognizes the significance of the technology in the current increasingly complex and volatile global security environment and its impact on national security. Rapid technological progress, partly driven by AI development will ultimately change the character of war (DoD, p. 2–3).

According to the strategy, the attainment of the US strategic defense objectives and the maintenance of US global influence will, among other things, require adopting a strategic approach, part of which will include rebuilding its force posture. The US forces should possess decisive advantages for any likely conflict, while remaining proficient across the entire spectrum of conflict. To that end, the strategy makes it imperative to modernize American forces' key capabilities. This will entail specific investments in critical areas, one of the eight modernization programs listed in the strategy being advanced autonomous systems. The DoD is planning to substantially invest in the military application of autonomy, artificial intelligence, and machine learning in order to gain competitive military advantages (DoD, 2018, p. 7).

The first key government document devoted specifically to AI, acknowledging its prominence for national security, was the Executive Order on Maintaining American Leadership in Artificial Intelligence issued by US President Trump on February 11, 2019. The document, being a declaration of US intent to expand and formalize its efforts to support AI development, sets forth the policy priorities, principles, objectives for the US to promote and protect its AI R&D for economic and societal development as well as national security objectives. It establishes six strategic goals for executive departments and agencies (WH, 2019, Sec. 2):

1. promotion of investment in AI R&D with industry, academia as well as international partners and allies;
2. improvement of access to high-quality and fully traceable federal data;
3. gradual removal of the barriers to wider AI application;
4. provision of technical standards to minimize vulnerability to cyberattacks;
5. training of future US AI researchers and users;
6. development of an action plan to protect the American advantage in AI.

5 Adoption of AI at DoD's Level

The following day, the Department of Defense published the summary of its 2018 classified artificial intelligence strategy: Harnessing AI to Advance Our Security and Prosperity. The strategy is a follow-on to the National Defense Strategy and complements DoD's efforts to modernize information technology to support the warfighter, defend against cyberattacks and leverage emerging technologies. As AI is actually ubiquitous, the strategy sees it necessary to employ its full potential to achieve national security goals. Hence, the aim of the document is twofold: to address the different challenges posed by AI as well as seize the opportunities offered by AI to advance security, preserve peace and stability in the long run.

The strategy provides a clear explanation of how the adoption of AI will benefit the DoD and the United States (DoD, 2019, p. 6). First, it will support and protect US service members and civilians around the world, for instance by employing AI employment for decisions-making processes, thereby reducing risk to ongoing operations and helping to lower the risk of civilian casualties and other collateral damage. It is also assumed that AI will help better maintain equipment, effect operational costs reduction or enhance readiness. Second, it will protect the United States and safeguard American citizens by providing increased protection and defense of American territory and/or US critical infrastructure from attack and disruption. Third, it will create an efficient and streamlined organization by making workflows simpler and more efficient, and certain tasks completed with greater speed and accuracy. Fourth, it will allow the US to become a pioneer in scaling AI across a global enterprise, as the DoD wants to be at the forefront of AI implementation for a variety of capabilities for other departments and agencies of the US government, but also coalition partners and allies. It hopes to establish the right approaches, standards and procedures as well as operational models.

The DoD's strategic approach for AI rests on the following five pillars (DoD, 2019, p. 7–8):

1. Delivering AI-enabled capabilities that address key missions;
2. Scaling AI's impact across DoD through a common foundation that enables decentralized development and experimentation;
3. Cultivating a leading AI workforce;
4. Engaging with commercial, academic, and international allies and partners;
5. Leading in military ethics and AI safety.

The DoD envisions the application of AI-enabled capabilities to enhance the decision-making process and key mission areas. It is expected that it will help to improve situational awareness and decision-making processes, increase the safety of operating equipment, make predicting maintenance needs of some pieces of equipment and supply demands more accurate and efficient, or, in general, streamline certain processes allowing one to reduce time spent on time-consuming, repetitive and somewhat unsophisticated tasks. AI systems that will be implemented should enhance military personnel capabilities by unburdening them of menial cognitive or physical tasks and making their work more effective (DoD, 2019, p. 11).

The DoD wants to encourage a bottom-up approach to accelerate the delivery and adoption of AI. This will mean fostering the development of AI solutions out in a decentralized manner, through experiments at the forward edge. For this to happen, the DoD plans to roll out a common infrastructure, consisting of platforms, procedures, standards, tools, services, etc., which will all make it possible to adapt and apply the solutions that have been worked out, speed up the experimentation and delivery of AI applications and help to promote successful AI prototypes (DoD, 2019, p. 7).

The absolutely crucial aspect to the implementation of the strategy is personnel. Since at present the DoD experiences a general shortage of qualified AI specialists, the DoD sees it imperative to develop the existing as well as acquire the new workforce with critical AI skills. The existing staff should be offered comprehensive skills development and career progression opportunities through dedicated programs allowing them to stay abreast of the developments in the field and acquire the necessary skills as well as knowledge. Equally significant will be to acquire world-class specialists as well as knowledge from outside to complement the existing personnel and to make sure that the AI development team is able to address the most pressing challenges (DoD, 2019, p. 14).

The authors of the strategy are cognizant of the fact that AI development on a global scale will not materialize solely within the confines of the US government, and as AI advances de facto have their origins outside the military, it is imperative to bridge the gap between the civilian and defense sectors with regard to AI. That is why, the document stresses the indispensability of forging strong partnerships across the whole process with academic institutions and commercial entities, which are at the forefront of modern AI advancement, as well as international partners and allies to create a community jointly facing the challenges. This approach should make sure the academia engages in research responding to national security goals and educate the next generation of AI workforce, leaders in the civilian AI industry understand and can contribute to tackling security challenges, and new individuals and novel ideas can be attracted to the DoD-driven AI ecosystem (DoD, 2019, p. 12–13).

Lastly, the strategy emphasizes the significance of AI development and employment in an ethical, humanitarian and safe manner. Hence, the US aspires to not only provide and follow guidelines in that regard, compliant with domestic law and upholding international standards, and work on reducing the risk of collateral damage, but also encourage that they are applied by other countries. The specific actions to be taken include developing AI principles for defense, investing in research and development for resilient, robust, reliable, and secure AI, promoting transparency in AI research, advocating for a global set of military AI guidelines, and using AI to reduce the risk of civilian casualties and other collateral damage (DoD, 2019, p. 15–16).

In order to expedite and streamline the development of AI-enabled capabilities across the DoD and beyond, the strategy has also made provision for the establishment of the Joint Artificial Intelligence Center (JAIC), which was brought into being in 2018. As might be expected the JAIC's specific tasks correspond to the strategy's pillars and include the delivery of AI-enabled solutions to address key missions, the establishment of a common foundation for scaling AI's impact across the DoD, the furtherance of AI planning, policy, governance, ethics, safety, cybersecurity, and multilateral coordination, and the gathering of the necessary talent (DoD, 2019, p. 9). The JAIC will be involved throughout the development of AI applications, becoming more focused on near-term execution and adoption. The center's work will complement the efforts of the Defense Advanced Research Projects Agency (DARPA), DoD laboratories, and other institutions involved in AI R&D.

The JAIC will provide AI capabilities within two distinct categories: National Mission Initiatives (NMIs) and Component Mission Initiatives (CMIs). The former are broad, joint projects run with a cross-functional team approach. The latter are component-specific projects solving a particular problem. They are run by other research organizations with the JAIC's support. The first NMIs initiated by the JAIC in early 2019 include Predictive Maintenance and Humanitarian Assistance and Disaster Relief (Joint Artificial Intelligence Center, n.d.). It should be mentioned at this point that DoD components can engage in AI R&D at their own discretion; nonetheless, they are obliged to coordinate with the JAIC any planned AI initiatives of $15 million or more annually (CRS, 2020, p. 9).

The actions to embrace AI for defense purposes have not been limited to the above only. Other efforts included (CRS, 2020, p. 5):

1. the publication of a strategic roadmap for AI development and delivery as well as the publication by the Defense Innovation Board of "AI Principles: Recommendations on the Ethical Use of Artificial Intelligence by the Department of Defense";
2. the establishment of the National Security Commission on Artificial Intelligence to conduct a comprehensive assessment of methods and means required to advance AI development for national security and defense purposes;

In addition, members of the US Congress have filed a number of bills addressing AI. They also organized the Congressional Artificial Intelligence Caucus to brief policymakers of the impacts of AI development and ensure that the US fully benefits from AI innovation (NSCAI, 2019, 21).

The US has been steadily increasing its AI funding. The DoD's unclassified expenditure grew more than fourfold, from $600 million in FY2016 to $2.5 billion in FY2021. The DoD has reported over 600 active AI projects (CRS, 2020, p. 2). Regarding the DoD's FY 2021 research and development budget, it is said to be the largest ever requested. Selected efforts include (Office of the Under Secretary of Defense (Comptroller)/Chief Financial Officer, 2020, p. 1–9):

1. Autonomy—Enhances speed of maneuver and lethality in contested environments; develops human/machine teaming (FY 2021, $1.7 billion);
2. Artificial Intelligence (AI)—Continues the AI pathfinders, Joint Artificial Intelligence Center (JAIC) and Project Maven (FY 2021, $0.8 billion).

True, the DoD's investment in AI has increased, but it has also been argued that additional outlays will be indispensable to keep step with America's competitors and avoid "innovation deficit" in military technology (NSCAI, 2019, p. 25).

Although the strategy documents, initiatives and measures have been adopted and implemented relatively recently, AI-related projects are already underway, and AI is being incorporated into a number of applications. The most notable example for intelligence, surveillance and reconnaissance purposes was Project Maven, used to identify hostile activity for targeting in Iraq and Syria. It used computer vision and machine learning algorithms to autonomously spot objects of interest in the footage obtained by UAVs (Vanian, 2018). In military logistics, AI has been employed for predictive maintenance, which makes it possible to tailor maintenance needs of a piece of hardware based on data analytics in lieu of standardized routine maintenance schedules. This solution has already been employed by the US Air Force in the F-35's Autonomic Logistics Information System and by the US Army in the Logistics Support Activity for its fleet of the Stryker combat vehicle (CRS, 2020, p. 10).

The US military is also exploring and testing AI's potential for cyberspace operations, information operations, command and control and its incorporation into the DoD's Joint All Domain Command and Control (JADC2) to create a common operating picture, as well as semiautonomous and autonomous vehicles under such programs as the Loyal Wingman (US Air Force), the Multi-Utility Tactical Transport (the Marine Corps), the Robotic Combat Vehicle (US Army), the Anti-Submarine Warfare Continuous Trail Unmanned Vessel (DARPA). The DoD is also researching other AI-facilitated capabilities to enable cooperative behavior, or swarming (cf. CRS, 2020, p. 9–15).

6   Conclusions

The research has endorsed the fact that AI provides numerous opportunities and challenges in a national security context. Its transformative potential has wide-ranging implications for the attainment of strategic security goals and global competition as such. There is a direct link between national security and technological development. Losing that technological edge may be detrimental to national interest. This reality has been recognized by many countries, including America's rivals, who have been investing heavily in AI research and development in for military use with a view to improving the combat readiness and decision-making of their forces, and delivering novel capabilities, further enhancing their dominance. At the same time, they are aggressively proclaiming their intention of becoming the global leader in AI. America faces true global competitors for military superiority.

With the situation being as it is, US policymakers have to prepare for the impacts of AI-related technologies. The question thus is: how does the US embrace AI for defense purposes so that it maintains its strategic position and is still capable of advancing security, peace and stability? It has been shown in the chapter that the US follows the same path as its competitors. It experiences the same challenges and problems. Even if somewhat belatedly, the US government has provided enhanced policy and guidance, stepping up investments in AI and autonomy to ensure that the US maintains the competitive military advantage over adversaries, and, more importantly, attempting to build a robust AI ecosystem. The success in AI adoption will then be largely determined by the integration of the individual components of that ecosystem, as, for instance, AI capabilities will be developed mostly by third parties and contractors. All of them have to click into place. It should not be forgotten that, all in all, the US has a few sure advantages in terms of AI technology, industry, and talents which can afford it an opportunity to maintain its lead and win the competition with its rivals for US security, stability, and prosperity (cf. Allison & Schmidt, 2020, p. 22–24; Castro et al., 2019, p. 2).

References

Allen, G., & Chan, T. (2017). *Artificial intelligence and national security.* A study on behalf of Dr. Jason Matheny, Director of the US Intelligence Advanced Research Projects Activity (IARPA). Belfer Center for Science and International Affairs, Harvard Kennedy School. https://www.belfercenter.org/sites/default/files/files/publication/AI%20NatSec%20-%20final.pdf

Allison, G., & Schmidt E. (2020). *Is China beating the U.S. to AI supremacy?* Avoiding Great Power War Project. Belfer Center for Science and International Affairs, Harvard Kennedy School. https://www.belfercenter.org/sites/default/files/2020-08/AISupremacy.pdf

Auslin, M. (2018). *Can the Pentagon win the AI arms race? Why the US is in danger of falling behind.* Foreign Affairs. Retrieved September 13, 2020, from https://www.foreignaffairs.com/articles/united-states/2018-10-19/can-pentagon-win-ai-arms-race

Barnes, T. (2018). *China tests army of tiny drone ships that can 'shark swarm' enemies during sea battles.* Independent. Retrieved July 26, 2020, from https://www.independent.co.uk/news/world/asia/china-drone-ships-unmanned-test-video-military-south-sea-shark-swarm-a8387626.html

Bendett, S. (2017a). *Should the US army fear Russia's killer robots?* The National Interest. Retrieved September 13, 2020, from https://nationalinterest.org/blog/the-buzz/should-the-us-army-fear-russias-killer-robots-23098

Bendett, S. (2017b). *Red robots rising: Behind the rapid development of Russian unmanned military systems.* The Strategy Bridge. Retrieved August 29, 2020, from https://thestrategybridge.org/the-bridge/2017/12/12/red-robots-rising-behind-the-rapid-development-of-russian-unmanned-military-systems

Bendett, S. (2018). *Here's how the Russian military is organizing to develop AI.* Defense One. Retrieved August 28, 2020, from https://www.defenseone.com/ideas/2018/07/russian-militarys-ai-development-roadmap/149900

Buchanan, B. (2020). *The US has AI competition all wrong. Computing power, not data, is the secret to tech dominance.* Foreign Affairs. Retrieved August 7, 2020, from https://www.foreignaffairs.com/articles/united-states/2020-08-07/us-has-ai-competition-all-wrong

Castro, D., McLaughlin, M., & Chivot, E. (2019). *Who is winning the AI race: China, the EU or the United States?* Center for Data Innovation. Retrieved February 1, 2021, from https://euagenda.eu/upload/publications/who-is-winning-the-ai-race-china-the-eu-or-the-united-states.pdf

China State Council. (2017). *A next generation artificial intelligence development plan* (New America, Trans.). https://www.newamerica.org/documents/1959/translation-fulltext-8.1.17.pdf

CNA. (2020). *Artificial intelligence in Russia.* Issue 7. CAN. https://www.cna.org/CNA_files/centers/CNA/sppp/rsp/newsletter/DOP-2020-U-027701-Final2.pdf

CRS. (2017). *Overview of artificial intelligence* (IF10608). CRS Report Prepared for Members and Committees of Congress. Congressional Research Service. https://crsreports.congress.gov/product/pdf/IF/IF10608

CRS. (2020). *Artificial intelligence and national security* (R45178). CRS Report Prepared for Members and Committees of Congress. Congressional Research Service. https://crsreports.con gress.gov/product/pdf/R/R45178/9

DARPA. (2020). *Media forensics (MediFor).* Defense Advanced Research Projects Agency. Retrieved on November 1, 2020, from https://www.darpa.mil/program/media-forensics

Davies, G. (2017). *Russia boasts its robot tank outperformed manned vehicles in recent tests as Putin looks to expand droid army.* Mailonline. Retrieved October 30, 2020, from https://www.dailym ail.co.uk/news/article-5062871/Russia-boasts-robot-tank-outperformed-manned-vehicles.html

DIB. (2019). *AI principles: Recommendations on the ethical use of artificial intelligence by the department of defense. Supporting document.* Defense Innovation Board. https://media.def ense.gov/2019/Oct/31/2002204459/-1/-1/0/DIB_AI_PRINCIPLES_SUPPORTING_DOCUME NT.PDF

DoD. (2018). *Summary of the 2018 national defense strategy.* Department of Defense. https://dod. defense.gov/Portals/1/Documents/pubs/2018-National-Defense-Strategy-Summary.pdf

DoD. (2019). *Summary of the 2018 department of defense artificial intelligence strategy: Harnessing AI to advance our security and prosperity.* Department of Defense. https://media.defense.gov/ 2019/Feb/12/2002088963/-1/-1/1/SUMMARY-OF-DOD-AI-STRATEGY.PDF

Gigova, R. (2017). *Who Vladimir Putin thinks will rule the world.* CNN. https://www.cnn.com/ 2017/09/01/world/putin-artificial-intelligence-will-rule-world/index.html

Groth, O. J., Nitzberg, M., & Zehr, D. (2019). *Comparison of national strategies to promote artificial intelligence, part 1.* Konrad Adenauer Stiftung. https://www.kas.de/documents/252 038/4521287/Comparison+of+National+Strategies+to+Promote+Artificial+Intelligence+Part+ 1.pdf/397fb700-0c6f-88b6-46be-2d50d7942b83?version=1.1&t=1560500570070

High-Level Expert Group on Artificial Intelligence. (2019). *A definition of AI: Main capabilities and scientific disciplines.* European Commission. https://www.aepd.es/sites/default/files/2019- 12/ai-definition.pdf

JAIC. (2020). *About the JAIC.* Joint Artificial Intelligence Center. Retrieved September 25, 2020, from https://www.ai.mil/about.html

Kallenborn, Z. (2019). *What if the U.S. military neglects AI? AI futures and U.S. incapacity.* War on the Rocks. Retrieved February 1, 2021 from https://warontherocks.com/2019/09/what-if-the- u-s-military-neglects-ai-ai-futures-and-u-s-incapacity

Kania, E. B. (2017). *Battlefield singularity: Artificial intelligence, military revolution, and China's future military power.* Center for a New American Security. https://s3.us-east-1.amazonaws. com/files.cnas.org/documents/Battlefield-Singularity-November-2017.pdf?mtime=201711292 35805&focal=none

Layton, R. (2018). *Algorithmic warfare. Applying artificial intelligence to warfighting.* Air Power Development Centre. https://www.researchgate.net/profile/Peter_Layton/publication/326248 009_Algorithmic_Warfare_Applying_Artificial_Intelligence_to_Warfighting/links/5b40ae1f0 f7e9bb59b112b93/Algorithmic-Warfare-Applying-Artificial-Intelligence-to-Warfighting.pdf

Markotkin, N., & Chernenko, E. (2020). *Developing artificial intelligence in Russia: Objectives and reality.* Carnegie Moscow Center. https://carnegie.ru/commentary/82422

New, J. (2018). *Why the United States needs a national artificial intelligence strategy and what it should look like.* Center for Data Innovation. http://www2.datainnovation.org/2018-national-ai- strategy.pdf

NSCAI. (2019). *Interim report. November 2019.* Washington: National Security Commission on Artificial Intelligence. https://www.nscai.gov/wp-content/uploads/2021/01/NSCAI-Interim-Rep ort-for-Congress_201911.pdf

Nurkin, T., Bedard, K., Clad, J., Scott, C., & Grevatt, J. (2018). *China's advanced weapons systems.* The US-China Economic and Security Review Commission. https://www.uscc.gov/sites/default/ files/Research/Jane%27s%20by%20IHS%20Markit_China%27s%20Advanced%20Weapons% 20Systems.pdf

Nurkin, T., & Rodriguez, S. (2019). *A candle in the dark: US national security strategy for artificial intelligence.* Atlantic Council Strategy Paper. Atlantic Council. https://www.atlanticcouncil.org/wp-content/uploads/2019/12/AC_CandleinDark120419_FINAL.pdf

Office of the President of the Russian Federation. (2019). *Decree of the President of the Russian Federation on the development of artificial intelligence in the Russian Federation* (Center for Security and Emerging Technology, Trans.). Moscow. https://cset.georgetown.edu/wp-content/uploads/Decree-of-the-President-of-the-Russian-Federation-on-the-Development-of-Artificial-Intelligence-in-the-Russian-Federation-.pdf

Office of the Under Secretary of Defense (Comptroller)/Chief Financial Officer. (2020). *Defense budget overview: United States department of defense FY2021 budget request.* Washington. Revised May 13, 2020. https://comptroller.defense.gov/Portals/45/Documents/defbudget/fy2021/fy2021_Budget_Request_Overview_Book.pdf

Pellerin, C. (2017). *Project Maven to deploy computer algorithms to war zone by year's end.* US Department of Defense News. Retrieved January 19, 2021, from https://www.defense.gov/Explore/News/Article/Article/1254719/project-maven-to-deploy-computer-algorithms-to-war-zone-by-years-end

Polyakova, A. (2018). *Weapons of the weak: Russia and AI-driven asymmetric warfare.* A Blueprint for the Future of AI: 2018–2019. Brookings. https://www.brookings.edu/research/weapons-of-the-weak-russia-and-ai-driven-asymmetric-warfare/

Scharre, P. (2016). *Autonomous weapons and operational risk.* Ethical Autonomy Project. Center for a New American Security. https://s3.us-east-1.amazonaws.com/files.cnas.org/documents/CNAS_Autonomous-weapons-operational-risk.pdf?mtime=20160906080515&focal=none

Schmidt, E., & Work, B. (2019). *The US is in danger of losing its global leadership in AI.* The Hill. Retrieved September 13, 2020, from https://thehill.com/blogs/congress-blog/technology/473273-the-us-is-in-danger-of-losing-its-global-leadership-in-ai

Seligman, L. (2018). *Pentagon's AI surge on track, despite google protest.* Foreign Policy. Retrieved January 20, 2021, from https://foreignpolicy.com/2018/06/29/google-protest-wont-stop-pentagons-a-i-revolution

Sheppard, L. R., Hunter, A. P., Karlén, R., & Balieiro, L. (2018). *Artificial intelligence and national security. The importance of the AI ecosystem.* A Report of the CSIS Defense-Industrial Initiatives Group. Center for Strategic and International Studies. https://csis-website-prod.s3.amazonaws.com/s3fs-public/publication/181102_AI_interior.pdf?6jofgIIR0rJ2qFc3.TCg8jQ8p.Mpc81X

Sherman, J. (2019). *Why the U.S. needs a national artificial intelligence strategy.* World Politics Review. Retrieved February 1, 2021, from https://www.worldpoliticsreview.com/articles/27642/why-the-u-s-needs-a-national-artificial-intelligence-strategy

Simonite, T. (2017). *For superpowers, artificial intelligence fuels new global arms race.* Wired. Retrieved September 6, 2020, from https://www.wired.com/story/for-superpowers-artificial-intelligence-fuels-new-global-arms-race

The White House. (2017). *National security strategy of the United States of America.* https://www.whitehouse.gov/wp-content/uploads/2017/12/NSS-Final-12-18-2017-0905.pdf

The White House. (2019). *Executive order on maintaining American leadership in artificial intelligence.* https://www.whitehouse.gov/presidential-actions/executive-order-maintaining-american-leadership-artificial-intelligence/

US Senate. (2016). *Hearing on the dawn of artificial intelligence* [114th Cong., 2nd sess.]. Subcommittee on Space, Science, and Competitiveness of the Committee on Commerce, Science, and Transportation. https://www.govinfo.gov/content/pkg/CHRG-114shrg24175/pdf/CHRG-114shrg24175.pdf

US Government Accountability Office. (2017). *Military acquisitions. DOD is taking steps to address challenges faced by certain companies* (GAO-17-644). Report to the Committee on Armed Services, US Senate. https://www.gao.gov/assets/690/686012.pdf

Vanian. J. (2018). *Defense department is using Google's AI tech to help with drone surveillance.* Fortune. Retrieved November 5, 2020, from https://fortune.com/2018/03/06/google-department-defense-drone-ai/

West, D. M., & Allen, J. R. (2018). *How artificial intelligence is transforming the world*. https://www.brookings.edu/research/how-artificial-intelligence-is-transforming-the-world

Mikołaj Kugler , Ph.D., Senior lecturer, General Tadeusz Kościuszko Military University of Land Forces, Wrocław, Poland. His areas of expertise include: international security, Poland's foreign and security policy, US foreign and security policy, Polish military involvement in peacekeeping and stabilization operations, the impact of new technologies on security.

A Roadmap to AI: An Insight from the Saudi Vision 2030

Saad Haj Bakry and Bandar A. Al. Saud

Abstract Artificial Intelligence (AI) is receiving increasing worldwide attention due to its potential relating to innovation and socio-economic growth and development. Several countries worldwide initiated national AI initiatives to explore AI for their own benefits, and to gain a competitive edge in AI for future development. The purpose of this chapter is to develop a generic national AI initiative framework that can be useful for countries concerned with planning for their future with AI. Against the backdrop of Bakry's wide-scope and structured STOPE (Strategy, Technology, Organization, People, and Environment) view, the elements of currently available key national AI initiatives are mapped.. The resulting framework is then used as a base for proposing an initial AI initiative for Saudi Arabia, considering responding to the requirements of its national development vision known as "KSA Vision 2030". It is hoped that the developed framework and its application would provide a roadmap for developing or improving national AI initiatives in various countries.

Keywords Artificial intelligence (AI) · Cyberworld · National AI initiatives · Saudi Arabia · The STOPE view

1 Introduction

Since the first "industrial revolution" started with the "steam engine" in the eighteenth century, knowledge generation and utilization accelerated, leading to a second, third, and fourth industrial revolutions, as illustrated in Fig. 1.

It is apparent that the field of Artificial Intelligence (AI) is currently a leading field in the fourth industrial revolution. It is receiving increasing worldwide attention due to its great economic development potential, and promising future lifestyle

S. H. Bakry (✉)
Department of Computer Engineering, King Saud University, Riyadh, Saudi Arabia
e-mail: shb@ksu.edu.sa

B. A. Al. Saud
Department of Electrical Engineering, King Saud University, Riyadh, Saudi Arabia
e-mail: bmashary@ksu.edu.sa

© The Author(s), under exclusive license to Springer Nature Switzerland AG 2021   201
A. Visvizi and M. Bodziany (eds.), *Artificial Intelligence and Its Contexts*, Advanced Sciences and Technologies for Security Applications,
https://doi.org/10.1007/978-3-030-88972-2_13

Fig. 1 The "Industrial
Revolutions" (WEF, 2017)

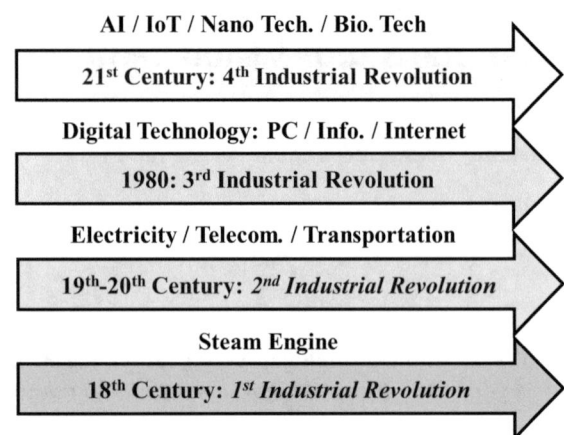

AI / IoT / Nano Tech. / Bio. Tech

21st Century: 4th Industrial Revolution

Digital Technology: PC / Info. / Internet

1980: 3rd Industrial Revolution

Electricity / Telecom. / Transportation

19th-20th Century: *2nd Industrial Revolution*

Steam Engine

18th Century: *1st Industrial Revolution*

(The Economist Intelligence.. 2016; 2018; Chatterjee, 2020; Troisi et al., 2021). In essence, AI is a digital technology associated with the Information and Communication Technology (ICT) and the Internet on the one hand, and with automation enabling things and machines to perform intelligent functions on the other. This enables such functions to be performed on the local level, and on wider levels that can be expanded up to the global space of the Internet (PWC, 2018; Paul et al., 2020).

The digital technology, with its current development in the fourth industrial revolution, can be viewed in multi-layer according to Fig. 2.

Wired and wireless broadband is the base layer that supports the Internet layer, and consequently the Internet of things, cloud computing and G5 (ITU-T Y. 2060 2012; ITU, 2012; Rodriguez, 2015). The applications of big data analytic receive data through that layer and is supported by AI (IBM 2018). In turn, AI can add intelligence to the layers below, in addition to its role of automation, where automated robots, with their sensors and mechanical functions, can operate and be controlled locally, or through the Internet and its various applications (Domingos, 2015). Of course, further progress is apparently underway in all digital technology tools and

Fig. 2 "AI" an integral part
of the cyberspace (The
authors)

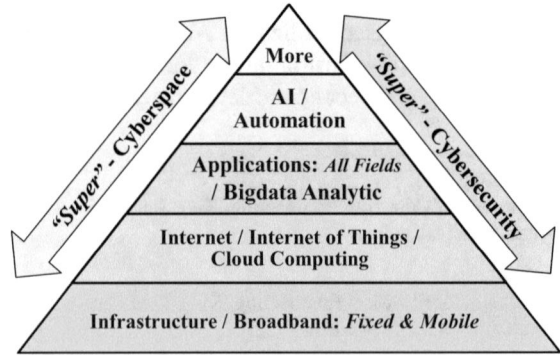

"*Super*" - Cyberspace

"*Super*" - Cybersecurity

More

AI /
Automation

Applications: *All Fields*
/ Bigdata Analytic

Internet / Internet of Things /
Cloud Computing

Infrastructure / Broadband: *Fixed & Mobile*

applications, particularly in AI (cf. Troisi et al., 2021). This raises several questions as to their impact on the society at large (Anastasiadou et al., 2021).

The word "cyberspace or cyberworld" is used to mean the world of ICT and the Internet. "Cyber" is originally a Greek word, which means "governance or control" (Cybernetics 2018). Like the physical world, the cyberworld needs protection and this is widely known as "cybersecurity" (Clark, 2010; ISO/IEC 2012). With AI, the "cyberworld" would become the "super-cyberworld", requiring "super-cybersecurity".

In response to the increasing importance of AI, some countries have recently initiated national AI initiatives to explore AI benefits and to gain a competitive edge in AI for future development (Dutton, 2018; MIT, 2019; China AI, 2018; Harhoff et al., 2018; AGDI, 2018; UAE, 2020; KSA, 2016). However, many countries have not yet provided enough attention to developing special AI initiatives, despite the fact that some of these countries have well-established national plans for future development. Here comes the purpose of this chapter, which is concerned with providing and using a generic national AI initiative framework that can be useful, as a guide, for countries concerned with developing or improving national AI plans for the future. For this purpose, the chapter addresses two main tasks.

- The first task is concerned with developing the targeted framework. It is based on an approach that maps the main issues of currently available key national AI initiatives on Bakry's wide-scope and structured STOPE view that involve the domains of: "Strategy, Technology, Organization, People, and Environment" (Bakry, 2004; Bakry & Alfantookh, 2012).
- The second task is associated using the resulting framework as a base for proposing an initial AI initiative for Saudi Arabia, considering responding to the requirements of its national development vision known as "KSA Vision 2030" (2016). This task involves two steps: understanding the country's vision, on the one hand; and developing the targeted AI initiative considering both the generic framework and Saudi vision, on the other.

2   The Generic Framework

This section is concerned with the first task of developing the targeted generic national AI initiative framework by mapping the main issues of available key national AI initiatives (Dutton, 2018; MIT, 2019; China AI, 2018; Harhoff et al., 2018; AGDI, 2018; UAE, 2020) on Bakry's wide-scope and structured STOPE view illustrated in Fig. 3.

Fig. 3 The five domains
"STOPE" framework
(Authors based on: Bakry,
2004; Bakry & Alfantookh,
2012)

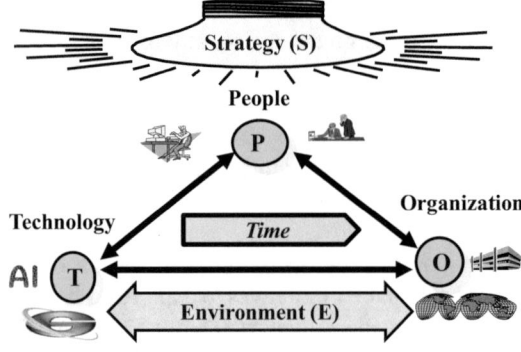

2.1 AI Strategic Issues

AI, together with its wide scope of applications, carry two main promises: the first is concerned with progress toward economic prosperity; and the second is associated with the improvement of the quality of life. Available key national AI initiatives give important "strategic development directions"; and these directions involve the following (Dutton, 2018; MIT, 2019; China AI, 2018; Harhoff et al., 2018; AGDI, 2018; UAE, 2020).

- Calling for "national dialogues" on AI; and for "national investment" in AI.
- Highlighting "research and development" in AI technology, and in its various fields of application.
- Considering building "AI industries" for producing: general purpose components; and special purpose systems.
- Underlining "AI use" in various fields, and emphasizing: economic development; improved social welfare; and enhanced public services to citizens and to businesses.

Table 1a gives the above directions against the countries that produced national AI initiatives. It should be noted here that among the national AI initiatives, the German initiative (Harhoff et al., 2018) stressed the need for looking at AI planning from an "ecosystem viewpoint". This means considering the expected interactions between AI and various elements associated with the national society, and may be with the whole world at large, on the one hand; and the environment conditions under which such interactions take place on the other.

2.2 AI Technology Issues

"Technology" issues addressed by key national AI initiatives are given in the following (Dutton, 2018; MIT, 2019; China AI, 2018; AGDI, 2018; UAE, 2020).

Table 1a Key AI national initiatives: (a): Strategic directions

Key directions		Key countries
General	Economic development	All
	Quality of life	
	Viewing AI as an ecosystem	Germany
Specific	Call for national dialogue	Japan/Poland/UAE
	Call for national investment	
	Research & development (R&D)	Australia/Canada/China/Germany/India/Mexico/Singapore/Sweden/Russia/Taiwan/UAE/UK/USA
	AI industries	China/France/India/Singapore/South Korea/Taiwan/UK/USA
	AI use: various applications and services	Denmark/Finland/Italy/Mexico/New Zealand/Singapore/UAE/USA

Source The Authors

- The development of "AI research" frameworks, plans, and roadmaps toward creativity and innovation in AI technology and its applications.
- Producing AI systems, including the following:

  - "AI general components" such as: sensors; neural networks chips; AI tools providing: machine learning, automated reasoning, natural language processing; AI process automation for driving robots; and other related components.
  - "Intelligent networked products", such as: vehicles; service robots; identification tools; and intelligent testing systems.
  - "AI special purpose products" for various "specific applications and services".

- Using AI in "intelligent manufacturing" for various types of production associated with different industries, as this would promote national productivity in various fields.
- Enhancing the "digital infrastructure", with AI capabilities for better services; this would include the following:

  - The Internet of things.
  - The G5 system and its promising applications.
  - Cybersecurity protection.
  - National big data analytic; and other potential enhancements.

- Using AI for various applications, including the following:

  - Public administration.
  - Urban solutions and transportation.
  - Education.
  - Health and health mobility.
  - Environment and sustainable development, such as clean water and renewable energy.
  - Financial solutions.
  - Military, defence and space.

Table 1b gives the above technology considerations against various key national AI initiatives that include them.

2.3 Organization Issues

"Organization" issues addressed by key national AI initiatives are given in the following (Dutton, 2018; MIT, 2019; China AI, 2018; Harhoff et al., 2018; AGDI, 2018; UAE, 2020).

- "AI leadership organizations", such as the following:

Table 1b Key AI national initiatives: (b): Technology issues

Key issues		Key countries
AI research and development (AI R&D)	Frames/Plans/Roadmaps	Australia/Canada/China/Germany/India/Mexico/Singapore/Sweden/Russia/Taiwan/UK/USA
	Producing AI systems	General AI components, such as: sensors; neural networks chips; tools providing: machine learning, automated reasoning, natural language processing; robotic process automation; and others	China/Asia's AI Agenda
	Intelligent and networked products	
	Special purpose products including robots	
AI in intelligent manufacturing	Promoting productivity in different industries	China/Japan/Taiwan /
AI in the digital infrastructure	Internet of Things (IoT)	Italy/Malaysia/Mexico/Poland/UK/USA
	G5 and applications	
	Cybersecurity protection	
	Bigdata analytic	
AI in various applications and services	Public administration	Italy/Mexico/Poland
	Urban solutions	Singapore/UAE
	Transportation	
	Education	Poland/Singapore/UAE
	Health and health mobility	France/Japan/Singapore/UAE
	Environment and sustainability	UAE
	Financial solutions	Singapore

(continued)

Table 1b (continued)

Key issues	Key countries
Military, defense and space	Russia/South Korea/UAE

Source The Authors

- – "A national strategic AI council" that would be involved in the following: an "AI forum" of people, universities, companies, and government departments brought together; "digital economy corporation"; and "AI government partnership with national technology companies".
  - – "A regional cooperation council", such as that of the Nordic-Baltic region.
  - – "An AI international innovation hub" such as that considered by Taiwan.
  - – "An advisory council on AI related ethics", such as that considered by Singapore.
- • "Special AI schools and colleges".
- • "AI research organizations", such as the following:

  - – Research institutes, including specialized research labs.
  - – Research clusters, with multi-disciplinary research teams.

- • "AI business organizations", such as:

  - – "AI business centers".
  - – "Marketing organizations".
  - – Support for "AI small-medium enterprises: SMEs".

Table 1c gives the above organization' issues against key national AI initiatives that include them.

2.4 People Issues

"People" issues addressed by key national AI initiatives are given in the following (Dutton, 2018; MIT, 2019; China AI, 2018; AGDI, 2018; UAE, 2020).

- • "Education", including the following:

  - – Basic education in "science, technology, engineering and mathematics: STEM".
  - – Basic "AI education for all".
  - – Advanced "AI education for specialized researchers and professionals".

- • "AI research scholarships".
- • "Incentives for attracting national and international AI talents".

Table 1d gives the above people's issues against the various national AI initiatives that include them.

Table 1c Key AI national initiatives: (c): Organization issues

Key issues		Key countries
AI leadership organizations	Strategic AI national council/AI National centers/Teams	Finland/Germany/Italy/Japan/Russia/Singapore
	AI forum	Malaysia/New Zealand
	Regional cooperation forum	Nordic-Baltic Region
	International Innovation hub	Taiwan
	AI ethics council	Singapore
Special AI schools	For AI education at different levels	China/Mexico/UK/South Africa/Singapore
AI research organizations	AI research institutes/Labs	Canada/France/Germany/Italy/Japan/Sweden/UK/USA
	AI science parks	
	AI government innovation agency	
	AI research clusters	
AI business organizations	AI business centers/with government cooperation	China/Finland/South Korea/Japan
	AI marketing organizations	India (Emphasizing the developing world)
	Support for AI SMEs/Start-ups	Denmark/France/Germany/Japan/South Korea

Source The Authors

Table 1d Key AI national Initiatives: (d): People issues (own studies)

Key issues		Key countries
Public awareness	Conceptual aspects and impact of AI	Malaysia/New Zealand
Education	STEM education	Canada/China/Mexico/Singapore/Singapore/UK/USA
	Basics of AI for all	
	AI for specialized professionals	
AI research	Scholarships	Australia/Canada
AI talents	Attracting national and international talents	China/Finland/Germany/India/Italy/Japan/New Zealand/Taiwan/USA

Source The Authors

2.5 Environment Issues

"Environment" issues addressed by key national AI initiatives are given in the following (Dutton, 2018; MIT, 2019; China AI, 2018; Harhoff et al., 2018; AGDI, 2018; UAE, 2020).

- "AI ethical related norms" expressed by a framework of ethics or ethical guidelines.
- "AI legal regulations", emphasizing the following:
  - Security and fairness.
  - Removing barriers on innovation and cooperation.
  - Making public and trusted data widely available.

- "AI technical standards", which would lead to:
  - Reliable AI products.
  - Safe and secure AI applications.
  - Efficient and effective use of AI.

- Encouraging "AI investment and funding".

Table 1e gives the above environment considerations against the various national AI initiatives that include them.

It should be noted here that the Italian initiative (AGDI, 2018) stressed the need for an ethical environment that considers AI to be a powerful tool in service of people for their benefits and not against them.

3 KSA Vision 2030

This section is concerned with the first step of the second task that is addressing the Kingdom of Saudi Arabia (KSA) Vision 2030 (KSA, 2016). As a member of the G20, KSA is one of the leading countries of the world. Figure 4 shows the map of KSA, and Table 2 gives World Bank data on KSA and on the world.

The country enjoys various advantages that provide strengths to its vision. These strengths include: having a geographical location that comes at "the heart of the Arabic and Islamic worlds"; being "a hub connecting three continents"; and enjoying wealth that makes it an "investment power house".

These strengths are illustrated in Fig. 5. Considering these strengths, the KSA vision has three main pillars, which views the future of KSA as: "a vibrant society"; "a "thriving economy"; and "an ambitious nation".

These pillars are illustrated in Fig. 6; and each of these three pillars is addressed in the following.

Table 1e Key AI national initiatives: (e): Environment issues (own studies)

Key issues		Key countries
AI ethical environment	AI ethical framework/guidelines	Australia/Canada/China/Finland/France/Germany/UK/India/Italy/Mexico/New Zealand/Singapore/Sweden/UK/USA
AI regulations	Security and fairness	Canada/USA/Denmark/France/Sweden/USA
	AI innovation and cooperation: removing barriers	USA
	Availability of trusted data	Japan/New Zealand/Taiwan/UK/USA
Technical standards	Reliable AI products	China/Nordic–Baltic Region
	Safe and secure AI applications	
	Efficient and effective use of AI	
Encouragement	AI investment	Japan/Poland/UAE

Source The Authors

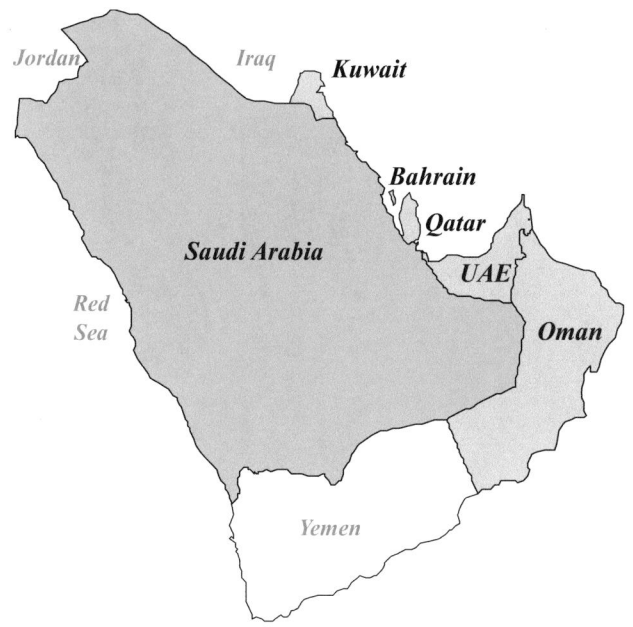

Source: The Authors

Fig. 4 KSA and the other GCC countries *Source* The Authors

Table 2 KSA and the world: key socio-economic indicators

ISSUE	KSA	The world	Unit
Area	2	130	Million km^2
Population	33	7530	Million
Pop. growth	2.03	1.158	(%) Annually
GDP per capita	20,761	10,714	US $
Proportion of natural wealth in GDP	27.2	1.894	(%)

Source The Authors based on World Bank

3.1 A Vibrant Society

This pillar has the following three sub-pillars, with each sub-pillar having directions and specific goals to be achieved by the year 2030 (KSA, 2016).

- The first sub-pillar is "strong roots", which has three directions and two main goals. Table 3a gives these directions and goals.

- The second sub-pillar is "fulfilling lives", which has four directions and three main goals. These are given in Table 3b.

Fig. 5 The strengths of the
KSA vision 2030. *Source*
(KSA, 2016)

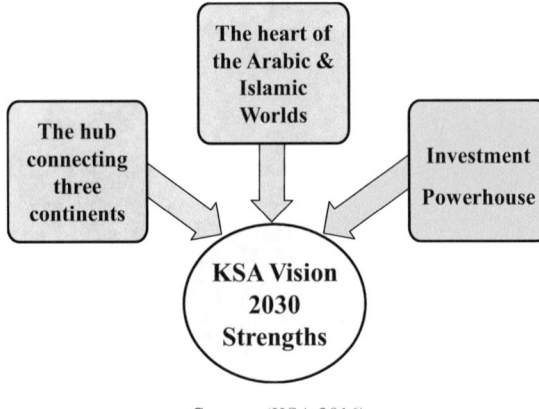

Source: (KSA 2016)

Fig. 6 The pillars of the
KSA vision 2030. *Source*
(KSA, 2016)

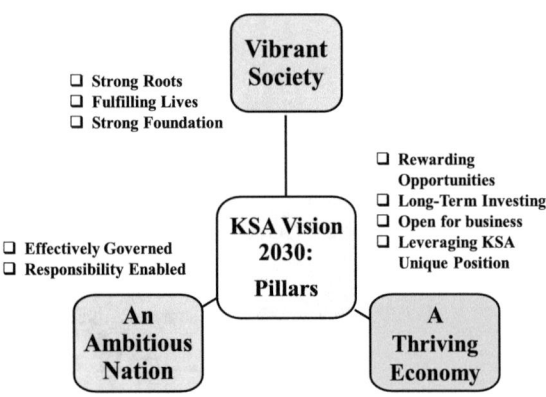

Source: (KSA 2016)

Table 3a KSA vision 2030: Pilar (1): Vibrant society (VS): (a): Strong roots: AI directions (KSA, 2016)

Directions	Goals
Living by Islamic law	• To increase KSA capacity to welcome Umrah visitors from 8 to 30 million every year
Focusing on efforts to serve UMRAH visitors	
Taking pride in national identity	• To more than double the number of KSA heritage sites registered with UNESCO

Source (KSA, 2016)

- The third sub-pillar is "strong foundation", which has four directions and two main goals. Table 3c gives these directions and goals.

Table 3b KSA vision 2030: Pilar (1): Vibrant society (VS): (b): With fulfilling lives: AI directions (KSA, 2016)

Directions	Goals
Promoting culture and entertainment	• To have three KSA cities be recognized in the top-ranked 100 cities in the world
Living healthy/being healthy	• To increase household spending on cultural and entertainment activities inside the KSA from the current level of 2.9 to 6%
Developing KSA cities	
Achieving environmental sustainability	• To increase the ratio of individuals exercising at least once a week from 13% of population to 40%

Source (KSA, 2016)

Table 3c KSA vision 2030: Pilar (1): Vibrant society (VS): (c): With strong foundation (KSA, 2016)

Directions	Goals
Caring for our families	• To raise KSA position from 26 to 10 in the Social Capital index
Developing our children's character	• To increase the average life expectancy from 74 to 80 years
Empowering our society	
Caring for our health	

Source (KSA, 2016)

3.2 A Thriving Economy

This pillar has the following four sub-pillars, with each sub-pillar having directions and specific goals to be achieved by 2030 (KSA, 2016).

- The first sub-pillar here is "rewarding opportunities", which has four directions and three main goals. Table 4a gives these directions and goals.
- The second sub-pillar is "investing for long term", which has three directions and three main goals. These are given in Table 4b.
- The third sub-pillar is "open for business", which has four directions and three main goals. Table 4c gives these directions and goals.

Table 4a KSA vision 2030: Pilar (2): Thriving economy (TE): (a): Rewarding opportunities (KSA, 2016)

Directions	Goals
Learning for working	• To lower the rate of unemployment from 11.6 to 7%
Boosting for our small businesses and productive families	• To increase SME contribution to GDP from 20 to 35%
Providing equal opportunities	• To increase women's participation in the workforce from 22 to 30%
Attracting talents, we need	

Source (KSA, 2016)

Table 4b KSA vision 2030: Pilar (2): Thriving economy (TE): (b): Investing for the long term (KSA, 2016)

Directions	Goals
Maximizing our investment capabilities	• To move from KSA current position as the 19th largest economy in the world into the top 15
Launching our promising sectors	• To increase the localization of oil and gas sectors from 40 to 75%
Privatising our government services	• To increase the Public Investment Fund's assets, from SAR 600 billion to over 7 trillion

Table 4c KSA vision 2030: Pilar (2): Thriving economy (TE): (c): Open for business (KSA, 2016)

Directions	Goals
Improving the business environment	• To rise from KSA current position of 25 to the top 10 countries on the Global Competitiveness Index
Rehabilitating economic cities	• To increase foreign direct investment from 3.8% to the international level of 5.7% of GDP
Establishing special zones	• To increase the private sector's contribution from 40 to 65% of GDP
Increasing competitiveness of our energy sector	

Table 4d KSA vision 2030: Pilar (2): Thriving economy (TE): (d): Leveraging KSA unique position (KSA, 2016)

Directions	Goals
Building a unique regional logistical hub	• To raise our global ranking in the Logistics Performance Index from 49 to 25 and ensure the Kingdom is a regional leader
Integrating regionally and internationally	• To raise the share of non-oil exports in non-oil GDP from 16 to 50%
Supporting our national companies	

- The fourth sub-pillar is "leveraging KSA unique position", which has three directions and two main goals. Table 4d gives these directions and goals.

3.3 An Ambitious Nation

This pillar has the following two sub-pillars, with each sub-pillar having directions and specific goals to be achieved by 2030 (KSA, 2016).

- The first sub-pillar here is "effective government", which has five directions and three main goals. Table 5a gives these directions and goals.

- The second sub-pillar is "responsibility enablement", which has three directions and three main goals. These are given in Table 5b.

Table 5a   KSA vision 2030: Pilar (3): An ambitious nation (AN): (a): Effectively governed (KSA, 2016)

Directions	Goals
Embracing transparency	• To increase non-oil government revenue from SAR 163 billion to SAR 1 Trillion, • To raise KSA ranking in the Government Effectiveness Index, from 80 to 20 • To raise our ranking on the E-Government Survey Index from our current position of 36 to be among the top five nations
Protecting our vital resources	
Engaging everyone	
Committing to efficient spending and balanced finances	
Organizing ourselves with agility	

Table 5b   KSA vision 2030: Pilar (3): An ambitious nation (AN): (b): Responsibly enabled (KSA, 2016)

Directions	Goals
Being responsible for our lives	• To increase household savings from 6 to 10% of total household income • To raise the non-profit sector's contribution to GDP from less than 1 to 5% • To rally one million volunteers per year (compared to 11,000 now)
Being responsible in business	
Being responsible to society,	

4   AI Initiative Based on KSA Vision 2030

This section is concerned with the second step of the second task that is developing a proposed KSA AI initiative. The initiative is based on the generic AI STOPE framework that considers key national AI initiatives (Dutton, 2018; MIT, 2019; China AI, 2018; AGDI, 2018; UAE, 2020), on the one hand; and on the development directions of the KSA vision 2030, on the other (KSA, 2016). It is described in the following according to the five domains of the generic AI STOPE framework.

4.1   KSA AI Strategic Directions

The "central AI strategic goal" would be to consider AI for enhancing advancement toward the achievement of the KSA vision 2030 that is directing AI efforts toward contributing to the following:

- The pillar of the "vibrant society (VS)", which involves: strong roots; fulfilling lives; and strong foundation.
- The pillar of the "thriving economy (TE)", which comprises of: rewarding opportunities; investing for long term; openness to business; and leveraging the unique position of the country.
- The pillar of the "ambitious nation (AN)", which includes effective government; and enabled responsibility.

4.2   KSA AI Technology Issues

The AI "technology domain" is the "core" of any national AI initiative. Table 6a considers key AI technology issues that can contribute to the development requirements of the KSA vision 2030. Further explanations are given in the following.

- AI "research and development" would be directed toward the needs of the three pillars of the KSA vision 2030: the VS; the TE; and the AN.

Table 6a "AI Technology Issues" for KSA vision 2030 (KSA, 2016)

Technology issues		KSA vision
Research and development	Supporting the various aspects of the three pillars of the KSA vision 2030 and beyond	VS/TE/AN
Producing/AI systems	General AI components	TE
	Systems for local use and local market	TE (see below)
	Systems for the regional market	TE
	Systems for the international market	
	For new business models (Employment)	TE: Rewarding opportunities
AI in industry/manufacturing	AI in industry/manufacturing in general	TE
	"Oil, Gas Sector"	TE: Long term investment
AI in digital infrastructure	Enhancing digital services	VS/TE/AN
	Creating new digital services	
	Enabling effectiveness and efficiency	
AI in applications and services	AI for "Umrah Services"	VS: Strong roots
	AI for "Smart Cities"	VS: Fulfilling lives
	AI for "Cultural, Entertainment Activities, and Exercise (Sport)"	
	AI for "Families and Raising Children"	VS: Strong foundation
	AI for "Healthcare"	
	"Public Services": Transparency and Performance	AN: Effectively governed
	"Security of Vital Resources"	
	"Social Services"	
	"Legal Services"	AN: Responsibility enabled

- "Producing and acquiring" AI technology systems that would contribute to the various requirements needed for TE. They would involve: general AI components; systems for local, regional, and international markets; and systems for new business models that while promoting productivity, they provide new human employment opportunities.
- Providing AI technology "support" to various industries, including those in the oil and gas sector as needed by the TE pillar.
- Enhancing the "digital infrastructure" with AI technology for the benefit of all pillars: VA, TE, and AN, as all are using the digital infrastructure.
- Exploring AI "capabilities" in applications and services that corresponds to the requirements of the pillars of: VS, such as "Umra services and smart cities"; TE, such as "building a unique regional logistical hub"; and AN, such as supporting effective governments and responsible citizens

4.3   KSA AI Organization Issues

The AI organization domain is concerned with identifying the potential organizations that would be "responsible" for the KSA AI initiative. These organization are outlined in Table 6b and are explained further in the following.

- AI "leadership" organizations that can be established at the national level, sector level, and task level. AI professional societies can also be considered among

Table 6b   "AI Organization Issues" for KSA vision 2030 (KSA, 2016)

Organization issues		KSA vision
AI leadership organizations	National level: Strategic development directions/ethics/regulations/standards/cooperation: different levels	Roadmap for AI technology generation, production and application toward contributing to: VS; TE; and AN
	Sector level/task level	
	Professional AI societies	
Media organizations	AI Knowledge for all	AI Knowledge for roadmap toward contributing to: VS; TE; and AN
AI schools and colleges	At all levels: General Education/Technical Education/University Education	
AI research institutes and labs	Within academic institutions	
	Special institutions	
AI government departments and business enterprises	Production of AI systems	Putting roadmap into action toward contributing to: VS; TE; and AN
	Applications and services	
	Operation and maintenance of AI systems	
	Support (e.g. Banks/Investment)	
	Emphasis SMEs	

these leadership organizations. The recommended top-level national organization would be responsible for initiating a "road map" for AI technology generation, production and utilization toward contributing to the pillars of the KSA vision.

- AI organizations concerned with "knowledge" would include: media organizations that would be mainly concerned with awareness; AI schools and colleges providing education and training; and AI research institutes and labs concerned with the generation of AI and AI related knowledge. These organizations would follow the AI road map toward the KSA vision and its pillars.
- AI government and "business" organizations that would be concerned with the production, acquisition and utilization of AI technology for different purposes associated with the pillars of the KSA vision.

4.4 KSA AI People Issues

All AI activities are "for people and by people". The AI people domain here is concerned with those who would be associated with all AI activities, including: using, supporting, running, designing and creating AI technology toward contributing to KSA vision pillars. For this purpose, AI programs for people would be needed; and these would include the following.

- AI "awareness" programs for all.
- AI "education and training" programs for AI specialists.
- AI "research and development" programs for competing with AI advancement and contributing to it.

In addition, attracting AI "talents" is always needed at the national level and at the international level. Potential talents of the gifted should be cared for and explored for future benefits. Table 6c provides a structured view of the suggested people domain.

4.5 KSA AI Environment Issues

The environment associated with certain activities either "make" them or "break" them. The environment needed for the AI roadmap toward contributing to KSA vision pillars should consider the following main issues.

- Maintaining "ethical behavior" by people, and this can be supported by: the awareness programs mentioned above; regulations and legal rules; in addition to special AI guarding systems.
- Developing and adopting "technical standards" for reliable and compatible AI technology. Such standards should include not only "specifications" of AI technologies, but also "best practices" in using them, and "indicators" for measuring quality and progress.

Table 6c "AI people issues" for KSA vision 2030 (KSA, 2016)

People issues		KSA vision
Awareness programs	Programs for dealing with AI illiteracy among all	Preparing the right people for the roadmap and for a long sustainable development toward contributing to: VS; TE; and AN and beyond
Education and training curriculum	STEM and AI basic knowledge programs: education at all levels	
	AI education for professionals	
Research and development support	General AI creativity and innovation	
	Directed AI creativity and innovation research scholarships	
Talent	Attracting national talents	
	Attracting international talents	
	Special care for the gifted and talented	

- Encouraging people and organizations to "perform" their work with enthusiasm to respond to the increasing competitiveness of the world.
- Supporting cooperating at all levels: local, national and international.

Table 6d provides an overall view of the suggested people domain.

Table 6d "AI Environment Issues" for KSA vision 2030 (KSA, 2016)

Environment issues		KSA vision
AI ethical environment	For ethical behavior	Ensuring the right environment for the AI roadmap and for a long sustainable development toward contributing to: VS; TE; and AN and beyond
AI regulations	Controlling behavior	
Technical standards	Should include AI technology specifications; best practices in using them; and indicators for measuring quality and progress	
Encouragement and cooperation	Support for AI production, application, and trade: (Business Zones/Energy Sector)	
	Cooperation: local/national/international (Regional Logistic Hub)	

5 Conclusions

This chapter has two main benefits. The first is the development of a generic structured national AI initiative framework based on Bakry's wide-scope view of "Strategy Technology, Organization, People and Environment: STOPE", considering the issues of available key AI national initiatives. The second is the use of the framework to propose an "AI initiative for KSA" considering the requirements of its Vision 2030. These two benefits together would hopefully lead to a third one that is the future use of the generic framework for the development of AI initiatives for other countries, or for improving existing ones.

In the development of the proposed "AI initiative for KSA", the KSA Vision's pillars of "a vibrant society; a thriving economy; and an ambitious nation" have been considered according to AI STOPE domains. The initiative considered the "AI technology" domain as the "core" for AI development in the country. The "AI organization" domain is considered to be associated with identifying the potential organizations that would be "responsible" for the initiative. The "AI people" domain viewed the KSA AI initiative to be "for people and by people"; and it considered various programs to promote their "capabilities and talents". The "environment" domain is considered the domain that "makes or breaks" the targeted AI initiative. It considered human behavior, technical standards, competitiveness and cooperation requirements.

It is finally hoped that the chapter would provide AI researchers and planners with a roadmap for developing and improving national AI initiatives in various countries.

References

AGDI. (2018). AI at the service of citizens, The Agency for Digital Italy, March.
Anastasiadou, M., Santos, V., & Montargil, F. (2021). Which technology to which challenge in democratic governance? An approach using design science research. *Transforming Government: People, Process and Policy.* https://doi.org/10.1108/TG-03-2020-0045
Bakry, S. H.; Alfantookh, A. (2012). Toward building the knowledge culture: reviews and a KC-STOPE with Six Sigma View, Chapter in the Book: Organizational Learning and Knowledge: Concepts, Methodologies, Tools and Applications, IGI Global, pp. 506–524, 2012.
Bakry, S. H. (2004). Development of e-Government: A STOPE View. *International Journal of Network Management, 14*(5), 339–350. https://www.google.com/search?q=Master+Algorithm&client=firefox-b-ab&source=lnms&tbm=vid&sa=X&ved=0ahUKEwjC-cbbgKTeAhWuzIUK HZMSB5MQ_AUIDygC&biw=958&bih=964.
Chatterjee, S. (2020). AI strategy of India: Policy framework, adoption challenges and actions for government. *Transforming Government: People, Process and Policy, 14*(5), 757–775. https://doi.org/10.1108/TG-05-2019-0031
China AI Development Report. 2018. China Institute for Science and Technology Policy at Tsinghua University, July 2018.
Clark, D. (2010). Characterizing cyberspace: past, present and future, MIT, CSAIL.
Domingos, P. (2015). The master algorithm, 28 Nov 2015—Uploaded from Talks at Google.
Dutton, T. (2018). An overview of National AI Strategies, June 28.

KSA. (2016). Vision 2030, Kingdom of Saudi Arabia, 2016. https://vision2030.gov.sa/en. Accessed March 2020.

Harhoff, D., Heumann, S., Jentzsch, N., Lorenz, P. (2018). Outline for a German strategy for artificial intelligence (2018). https://medium.com/politics-ai/an-overview-of-national-ai-strategies-2a70ec6edfd. Accessed January 2020.

IBM, Big data analytics. https://www.ibm.com/analytics/hadoop/big-data-analytics. Accessed December 2018

ISO/IEC 27032. (2012). Guidelines for Cybersecurity, Information Technology-Security Techniques. 2012.

ITU. (2012). Focus group on cloud computing, technical report. Part 1: Introduction to the cloud ecosystem: definitions, taxonomies, use cases and high-level requirements, 2012.

ITU-T Y.2060. (2012). Overview of the internet of things, global information infrastructure: Internet protocol aspects and next generation networks, Switzerland, 2012.

MIT. (2019). Technology review: Insight/Research, Asia's AI agenda: AI for business, MIT, 2019.

Paul, M., Upadhyay, P., & Dwivedi, Y. K. (2020). Roadmap to digitalisation of an emerging economy: A viewpoint. *Transforming Government: People, Process and Policy, 14*(3), 401–415. https://doi.org/10.1108/TG-03-2020-0054

PWC. (2018). AI Predictions: 8 insights shape business strategy, Price Waterhouse Cooper: pwc.com/us/AI2018.

Rodriguez, J. (Ed.). (2015). *Fundamentals of 5G mobile networks*. Wiley.

The Economist Intelligence Unit, Artificial Intelligence in the Real World: The business case takes shape, A Report from The Economist Intelligence Unit Limited, 2016.

Troisi, O., Visvizi, A., & Grimaldi, M. (2021). The different shades of innovation emergence in smart service systems: The case of Italian cluster for aerospace technology. *Journal of Business & Industrial Marketing*. https://doi.org/10.1108/JBIM-02-2020-0091

UAE. United Arab Emirates Artificial Intelligence Strategy: http://www.uaeai.ae/en/. Accessed March 2020.

The Economist Intelligence Unit, Intelligent Economies: AI's transformation of industry and society, A Report from The Economist Intelligence Unit Limited, 2018.

WEF. (2017). Realizing Human Potential in the Fourth Industrial Revolution: An Agenda for Leaders to Shape the Future of Education, Gender and Work, White Paper, World Economic Forum, Geneva, 2017.

Saad Haj Bakry is Professor at the Department of Computer Engineering, King Saud University, Saudi Arabia. He received his PhD from Aston University, England; and he is a Chartered Engineer and a Fellow of the Institution of Engineering and Technology. He has around 200 technical publications, including 14 books. He was also involved in the development of various national Saudi development plans. He has a weekly column on "knowledge and development" in "Aleqtisadia" Arabic daily. His e-mail address is: shb@ksu.edu.sa.

Bandar A. Al-Saud is Associate Professor at the Department of Electrical Engineering, King Saud University, Saudi Arabia. He received his PhD from University of Pittsburgh, USA. He has published over 25 refereed articles. He was a VP at Elm Company and was the Director of the National Information Center. In 2014, he was appointed as Assistant to the Minister of Interior for Technology Affairs, where he is still in this position. He is also on the board of King Abdulaziz and his Companions for Giftedness and Creativity, and Chairman of E-transactions at Ministry of Interior. His e-mail address is: bmashary@ksu.edu.sa.

Artificial Intelligence (AI) and International Trade

Katarzyna Żukrowska●

Abstract Artificial intelligence (AI) will have a profound impact on the economy and trade. This chapter examines the scope of changes AI will thus trigger in international trade. This chapter concentrates on the issue as it pertains to the European Union (EU) member states, the changes on the EU internal market, the EU's relations with outside partners, and as set within a wider context. The analysis was conducted amid the conditions caused by the slowing economy of 2018 and the onset of the COVID-19 pandemic. Crises often create new conditions and accelerate changes, which is supported by arguments from different fields. This chapter suggests that an acceleration of change had taken place and was caused by policies applied to overcome the pandemic and by changes in market players' behaviours. This chapter addresses three questions: Will the volume of trade expand or shrink in the nearest future? What trends can be expected in different groups of states using the criteria of advancement of development? Is AI a lever that stimulates the evolution of international trade?

Keywords International trade · Crisis and recovery · Industrial revolution 4.0 and trade · Digitalisation and trade · AI and trade; COVID-19 and trade

1 Introduction

Artificial intelligence (AI) is a new factor impacting various areas of the economy: the labour market, education system, production, and supply and demand. This chapter focuses on expected changes in international trade as a result of the differences in the readiness of specific groups of countries and their economies to use AI on a wider scale, the areas in which AI can be used, and the consequence on international trade. While this may be a broad approach, in this paper the problem is limited to the experiences of the EU, which can be seen as material to drawing much wider

K. Żukrowska (✉)
SGH Warsaw School of Economics, Warsaw, Poland
e-mail: kzukro@sgh.waw.pl

conclusions. The EU of 27 member states[1] is a big and important market in the world
economy and the national segments of this market vary in their level of integration.
Also, the EU has a huge network of partners (65) who enjoy preferential access
to the EU market (European Commission, 2020a). The objective of this chapter is
to examine whether (1) the volume of trade will expand or shrink in the nearest
future as a result of AI, (2) what trends can be expected in different groups of states,
taking into account their level of development; and, (3) can AI be delineated as a
stimulus helping to overcome the existing development gap between developed and
developing economies? The argument in this paper is structured as follows. The
first part concentrates on the results of trade in the last few years as affected by the
economic slowdown and the outbreak of the COVID-19 pandemic. The second part
offers prognoses for specified groups of countries, taking into account their level of
development and engagement in the international division of labour, specialisation,
and international trade.

2   Digitalisation and Artificial Intelligence

A basic definition of artificial intelligence (AI) is the ability of computerised machines
or digital controllers, or robots, to accomplish tasks that usually are performed by
human beings (Burkov, 2019; Goodfellow, 2016; Russer & Norvig, 2020; Visvizi,
2021). AI is used in reference to electronic systems equipped with software enabling
the collection and analysis of information, drawing conclusions, and undertaking
decisions. AI-based solutions have been applied in, for example, chess gameplay,
analysis of capital markets and decisions concerning the profitability of investments
and other.

The definition above, however, forces a distinction be made between digitalisation
and AI. In the first case, changes in production are shaped by information delivered
as a process of changing data into digital form, which can be easily read, stored,
and processed by a computer. Some companies still lag in their full use or capacity
of digital technologies. Digitalisation creates new possibilities to modify the business model. The main goal that drives businesses in this field is creating value from
new, advanced technologies by exploiting digital network dynamics stemming from
digitalisation. On top of that foundation, comes a new phase of digitalisation linked
with AI. This new stage goes beyond the process of the digital flow of information, i.e. it signals entry into a stage that until this moment was limited to human
decision-making. The novelty is that the collected data can be analysed in detail and
a computer-guided robot will propose the best possible solution given specific conditions and circumstances. The economic effectiveness of such decisions depends on
the collected and processed data and the formulation of the task for AI to solve.

Digital technologies reshape consumers' habits, which can be seen in the blurring
divide between traditional and online shopping. Digital technologies enable easier

---

[1] Does not include the UK, which formally exited the EU on 31 December 2020.

access to a growing number of suppliers while increasing suppliers' opportunities to win over the attention of a bigger group of consumers to their products. Digital technologies result in increasing the ability to enter a market, as well as in expanding the diversity of supply. Both have an impact on the intensification of competition, stimulate innovation, and help control prices.

Decisions concerning supplies of final products, spare parts, and components is proceeded by people have worked in the relevant capacity for a certain period of time. This means that they follow certain patterns of behaviour, habits, ties, relations, etc. established and developed in the past. This means that newly established suppliers and recently concluded agreements that can have an impact on costs and institutional competitiveness are not always taken into account. AI can bring important changes here, accelerated by the pandemic conditions. A state that offers intervention tries to support companies, businesses, and overall economic and financial activity (Balmford et al., 2020). The current difficulties and responding financial support, both occurring simultaneously, create new conditions in which companies can restructure their production, delivery chains, and finally, export markets. In practice, this means that available financial sources as well as long-lasting plans concerning production, competitiveness, and the profitability of the conducted activity can result in new investments on a larger scale than assumed today, because of the use of digital technologies. These technologies can be applied in different production stages, starting with product design, the physical production, analysis of supply chains and sales markets, followed by servicing.

Such changes can result in some cases in a return of industrial production to the group of countries labelled as "developed" or even in "the post-industrial stage of development". The trendy process of "reindustrialisation" means some production lines can return to markets from which industry has been leaving continuously since the 1970s. This return, however, is not the same conditions as in the past. The first difference is in the number and kind of workers employed in factories who often worked on production lines. Today, most will be replaced by robots. A number of products, forms, and components can be replaced by 3D or "4D" printing technology. The return of industrial production and scale will be limited by the—expanding—opportunities of B2B (business-to-business), B2H (business-to-household), H2H (household-to-household) deliveries. These can be achieved on both the national and international market, depending on the product type and access to information concerning supplies (cf. Ciuriak & Ptashkina, 2018). All this is enabled by expanding the presence of the internet as a supply channel (Edwards, 2013).

These dynamically developing technologies mean, for instance, that the emerging technique of 4D printing is based on 3D printing technology but with time and in response to data, can change shape or move, imitating natural processes. This can have a revolutionary impact on production, construction, medicine, space exploration, etc. Companies engaged in this research include Samsung Electronics, Sony, Dassault Systems, Google, Hexagon, Dreamworks, Autodesk, Stratasys, 3D Systems Corporation, Faro Technologies, Barco NV, Cognex Corporation, Dolby Laboratories, and others (BusinessWire, 2019). This revolution requires investments in infrastructure, including electrical systems, internet access, telecommunications, followed by advancement in education to prepare people for new roles and reality, defined by new and quickly changing conditions. The main drivers of such change will have consequences on employment, production, and trade. This will stimulate demand in some areas and reduce it in others.

3 Institutional Solutions Enhancing Trade in the Era of Digitalisation

The impact of digital technologies on trade depends on the institutional and legal solutions that frame related activities conducted at the international level. Further, there is a need to detail the influence of the General Agreement on Tariffs and Trade/World Trade Organisation (GATT/WTO) or others, such as the Organization of Economic Co-operation and Development (OECD), the EU, the European Free Trade Association (EFTA), or the European Economic Area (EEA), the Association of Southeast Asian Nations (ASEAN), the Southern Common Market (Mercosur), etc., on trade in which the trading partners engage in digital technologies (cf. Pauletto, 2020). It is assumed that WTO rules are technologically neutral, which means that solutions introduced in the Uruguay Round refer to goods, not to the role of channels used to conclude transactions or delivery. The same approach can be found in the solutions introduced by the General Agreement on Trade in Services (GATS). Conditions introduced by regional trade agreements, all approved by the GATT/WTO, have developed models of specified agreements that vary in structure, scope, scale, and depth. This means that all such solutions brought by international regimes can be more or less supportive in developing trade with the use of digital technologies, including AI. All depend on the scale of engagement of a country in international trade, its level of development, and its ability to cope with the new conditions and skills to use new technologies. The development stage of a country and policies applied by the authorities can bring it closer to the capacity to use these technologies, creating conditions for inclusion and helping to close the gap between insiders and outsiders (the latter, often self-fulfilling). In short, inclusion results from policies that bring economies into closer cooperation within the international system of the division of labour, with exclusion often stemming from inefficient policies supporting both anti-export, or pro-export types of production. This approach did not work in the past, and it will not work in the current conditions.

Such trade is not limited to fields traditionally perceived as linked with a specific digital trade regime. Digitalisation of trade has expanded beyond channels seen as suitable and into areas of the market in which supply and demand are shaped. Barriers to traditional trade also apply to digital trade, with the converse also true: elimination of conventional trade barriers stimulates digital trade. One idea within this framework is to create a single digital European market (CEF 2, 2020). Digital trade is seen as a bridge or short cut to the transfer of goods (WTO framework with Committee on Trade Development), services (the WHO's Council of Trade

Services), and intellectual property rights (Trade Related Intellectual Property Rights of the GATT/WTO) (the WTO's TRIPS) (WTO, 2018; WTO, 2020c). The WTO's rules and other regulations introduced on the global level create the framework for general trade for the 164 members and observer countries, as well as for international organisations that follow the WTO's policies and trade liberalisation moves.

Along with the WTO, international agreements form specific conditions easing access to the markets of the organisation's participants. Depending on the number of countries involved in such arrangements, they can be bilateral, multilateral, or global. Liberalisation of trade also can be introduced unilaterally. The latter solution does not lead to reciprocity, but it can lead to the application of the Most Favoured Nation Clause (MFNC) by partners. This is a step towards new solutions that start the process of opening the market. It must be emphasised that protectionist measures are often instruments in negotiations, helping to obtain better conditions. When a country tries to attract investors or partners to a specific market and has limited advantages to join other markets, opening its market can proceed as an autonomous move. This concerns the trade of goods and services and the flow of foreign direct investment (FDI). When a country has limited or no potential to export goods, services, or FDI, it is more interested in attracting inward flows in these areas. This can be accelerated by lowering the entrance barriers to the country's market without reciprocity.

Solutions applied in the WTO and regional free trade agreements (FTAs) show that the liberalisation of trade traditionally was limited to the "four freedoms": movement of goods, services, capital, and people (Greenway et al., 2013). In the EU, these freedoms were used to form the single market, called the "internal market". Some of the freedoms were easier to introduce, while others were seen as more difficult in the practical process of the elimination of barriers. This was also the case with services, which until now are not fully liberalised in Europe. The solutions applied in the EU's internal market are expanded to r-EFTA economies ("rest of" EFTA). The integrated structure that covers EU members and EFTA partners is called the European Economic Area (EEA). Due to recent advances in information and communication technology (cf. Visvizi & Lytras, 2019; Visvizi et al., 2021) the traditional 'four freedoms' ought to include free movement of digital information or streams of information (Lytras & Visvizi, 2021); certainly, caveats exist (cf. Calzada & Almirall, 2020).

Looking at data illustrating and comparing the share of merchandise export by region and their changes between 1980 and 2019—a period of 39 years—we can conclude that intensification of trade is higher in markets that are more open and integrated like the EU, Asian states, and North America. In the case of those countries, we can observe a fall in these shares, while in the case of less-advanced markets we can observe a slow rise in their shares (see Tables 1 and 2).

Interpretation of this data is simple and shows that more economic openness increases participation in trade and results in growth and development. Falling shares of regional trade in the so-called "developed North" are resulting from the growth in the share of trade with countries from other regions. This stems from the expansion of global value chains (GVC). This reveals that the observed tendencies in trade expansion create new conditions that promote trade liberalisation. Traditional arguments

Table 1 World merchandise export by region in 1983 and 2016, in percent share

Region	Share in 1980 (%)	Share in 2019 (%)
Europe	44.0	38.0
Asia	18.0	34.0
North America	17.0	14.0
Middle East	7.5	5.0
South and Central America and Caribbean	3.0	4.0
Africa	2.5	4.0

Source WTO (2020c) World Trade Report: Government policies to promote innovation in the digital age, Geneva: World Trade Organization (WTO), https://www.wto.org/english/res_e/publications_e/wtr20_e.htm

Table 2 Annual trade growth in the EU by partners in 2018–2019 (%)

Contents	All partners	FTA partners	Non-FTA partners
Imports	1.4	2.6	1.0
Exports	3.5	4.1	3.2
Total trade	2.5	3.4	2.1

Source European Commission (2019) European Annual report on implementation of EU trade agreements in 2019, EC, https://ec.europa.eu/commission/presscorner/detail/en/IP_19_6074 (12.11.2020)

used by economists to prove the necessity of using protectionist measures thus lose significance. We are observing more trade in the world economy than ever before, despite the drop in economic activity caused by the pandemic. We also can observe a growing number of countries active in trade as both importers and exporters. The goods being traded are far from the traditional industrial goods, fossil fuels, food and other agriculture products. The international division of labour no longer reflects a clear divide between industrialised and non-industrialised states. This reflects the replacement of an inter-branch division of labour by an intra-branch division of labour. In the industrial element of international trade, the share of components expands at a cost to finished goods trade. The net of institutional solutions and trade regulations that frame the conditions of trade becomes more and more complicated with the development of regional and bilateral FTAs. Finally, trade agreements that set conditions of trade between signatories seem outdated. Their obsolescence results from technical advancement, changes in economies and their structures, as well as changes in trade subjects.

4 Specific Conditions and Features Characterising the Crisis Caused by the COVID-19 Pandemic

The list of the conditions and observed features that characterise the current crisis is long and goes beyond the framework of this analysis (cf. Polese et al., 2021; Sokołowski, 2021; Visvizi & Lytras, 2020). A brief itemisation of features and conditions should be sufficient to show the debt, scale, and size of the crisis and these components' impact on the economy and applied policies. To address the problem formulated in the title of this article and expanded on in the introduction, one has to concentrate on interlinkages, such as those in: (1) production; (2) trade; (3) the possibility of introducing changes. This should help explain the fact that the listed features concentrate on the main problems comprising the background for reducing trade and further changing its directions and scale. To concentrate on these areas, there is the need to show some characteristics that can be divided into: (1) national features and solutions; (2) international features and solutions. Such a distinction seems to be well founded despite the fact that many countries followed a more or less similar model in their reactions but differed in the details of the applied policies differed. Also, the costs of the applied policies varied, as did the effectiveness and burdens with which such policies brought to the economies of specific markets.

Moreover, this crisis, like previous ones, has had a strong impact on economy and trade, but at the same time, the specificity of this crisis requires solutions that have a direct impact on the behaviour of consumers, producers, sellers, importers, and others. The solutions also influence the conditions shaping the institutional and legal frameworks of trade between countries even if they participate in integrated structures, use a common currency, or are part of an internal market with the four freedoms and subject to common trade policy. The latter concerns relations with third countries, some of which have free access to the EU market through signed free trade agreements. In this context, even less expected were solutions concerning closing borders to most people within the internal market. Also, solutions not applied in earlier crises also had effect on trade. Past crises, including the Depression of the 1930s, had an impact on production and trade and awakened nationalisms. Economic crises also affect politics, as demonstrated easily over the last century. All of the applied solutions eliminate some national tools of protection and are seen as building an irreversible status quo on liberal mechanisms. Regions and national markets are integrated on different scales, scopes, and depths, the intensity of ties, forming applied solutions that have an impact on trade policy in external relations. In the same period, for the first time in the history of integrated countries and economies, the applied solutions suspended part of the achieved results and were aimed at reaching deep homogeneity within the group (elimination of the virus).

Among others, the deepest effects of this can be seen within the EU states, and especially spotted among specific groups of states, such as the developed, emerging, developing, and least developed groups. New barriers or increased barriers to trade were observed in all the mentioned groups. The scale of the increase in barriers varied. In general, we can say that developed economies have relatively low tariff barriers, nevertheless, in most of these markets we can find non-tariff and para-tariff barriers, which are much more developed than in the rest of the world. Protection was also observed in emerging markets. Developing economies applied strongly diversified policies here. In most cases, the countries grouped here practice strong tariff protection. The least-developed economies apply in their trade policy more tariff tools of protection at the cost of non-tariff protection measures. These economies' reaction to the new crisis and policies applied by their more developed partners was most neutral. The economies listed in this specific group are not strongly engaged in the international division of labour, which results in their relatively limited reaction expressed as a rise in protectionism. Like in the case of the 2008 financial crisis, this group will likely receive an increased share of world flows of capital, goods, and services (Żukrowska, 2013).

The COVID-19 crisis has had the following implications:

1. Domestically:

    - fall in sales and production, followed by a drop in employment and GDP;
    - fall in budget revenues, increase in expenditures resulting in a higher budget deficit to GDP ratio, as well as increased public debt;
    - application of lockdown policies or avoiding total lockdown;
    - in post-industrial economies, a strong hit to the service sector (tourism, transport, trade, etc.);
    - awakening populist movements (protectionism, economic patriotism, demands for the return of industry, etc.).

2. In international relations:

    - Closing borders to the movement of people, resulting in closures of air trans-port connections, limiting tourism and related services (hotels, restaurants, museums, cinemas, theatres, namely all services that result in gatherings, especially large ones like festivals, etc.).
    - application of non-tariff and para-tariff protection, where applicable.
    - tensions also in established groupings of integrated structures or international organisations, as well as groups that help coordinate policies, applied tools, and undertaken actions), followed by trade wars that have had direct and indirect effects.
    - weakening of currency exchange rates, which has impacted price compet-itiveness (higher for exported goods and services and lower for imported goods and services).

The WTO forecasts for 2020 foresee a 9.2% decline in the volume of trade (WTO, 2020c). Despite the fact that all forecasts are uncertain, the current data collected by the WTO shows that the decline was not as severe as expected. Moreover, forecasts for this year (2021) show recovery in the volume of trade. There are branches and groups of countries that have suffered more and those that have suffered less or even gained. Although all mentioned occurrences took place in most economies, their scale differed by country and market. Nevertheless, states that have a big share of services and intellectual property in their exports did not suffer much from the application of tariff barriers by their trade partners. The application of tariff barriers in such a case increases prices on the market and limits competition. Reciprocity in such cases acts asymmetrically.

Diagnosis of the current economic situation seems to be the easiest part of the evaluation, which has to lead to a prescription that can bring the desired results. Some hints concerning what has to be done can be found in reviews of current policies applied by the EU and individual countries or markets.

This is illustrated by the EU's digital trade initiative, launched in 2017 (COM, 2017). This policy insists that states must meet specific targets on effectiveness, transparency, and values. It also is pointed at the stimulation of global trade, which is perceived as an opportunity for openness to flows of goods, services, people, and capital, all of which will fuel growth in the EU, enhancing competitiveness and consumer welfare.

The applied policy should be pointed at elimination of trade barriers. This task was also fulfilled by the European Commission, which conducted studies showing market access barriers for EU companies to third markets. This resulted in negotiations of new trade conditions by lowering the protection applied in specific areas and markets. Cohesion is seen as a method that can help globalisation work, but this, according to the EU Commission's evaluation, requires strengthening global governance, which has to be accomplished in all areas, including climate change. The above-mentioned ambitious goals can be accomplished under the condition of deep reforms in main member state activities that include education, investment, innovation, energy, fiscal and social policies, etc. The enumerated tasks seem today, in this period of pandemic, even more challenging then they were in 2017, which stems from the slowdown in economic activity leading to drops in GDP, followed by a reduction of budget revenues and an increase in expenditures. The response to the pandemic has enlarged the budget deficit of most member states, which increases the difficulty in overcoming the crisis, eliminating the negative occurrences caused by COVID-19, while still meeting the goals drafted in the digital strategy. This, in turn, is followed by recovery and interventionist measures by the EU.

The EU has furnished financial tools with some of its new financial facilities and is seeking other instruments, which in some cases seem to be advanced. All are aimed at increasing the ability of the European economy to respond and adjust to changes. Works are underway to make the European Globalisation Fund more flexible, as it is a solution that can be used immediately in case of a company closure or when quick responses to unexpected shocks in production chains take place. The available financial sources within the EU budget in the last year of the former Multiannual Financial Framework (MFF 2016–2020) and in the new MFF beyond 2020, as well as financial tools available outside the EU General Budget are also seen as sources that can meet two goals at one time: (1) solve the occurring problems; (2) introducing at the same time solutions incorporating innovative technologies.

Similar priorities are also listed in the context of the use of resources available within European Structural and Investment Funds (Goldstein, 2020), which also can be seen as additional financial tools enabling firm stabilisation of a company and introducing innovative technological changes at the same time. Such an approach seems to be important in light of the limited competitiveness of a number of European companies right now, even though they functioned for a while with the support of state interventionist measures.

The MFF 2021–2025 includes the Next Generation EU facility, introduced as a temporary financial instrument designed to stimulate the post-pandemic economic recovery. The money available in the coming MFF is evaluated as the biggest stimulus package ever covered by an EU budget—a total of €1.8 trillion. Besides stimulus, there is funding for companies to incorporate innovative technologies and enhance their competitiveness (Recovery Plan, 2020). The new MFF is constructed to maximise flexibility to finance needs not foreseen during the planning period. Such solutions are aimed at addressing the needs caused by the pandemic and the economic crisis that stems from it and future unknown but probable events (e.g., migration, pandemic, etc.). The fund promotes innovation, stimulates enhancement of the resilience of local economic systems, and coping with adjustments in the labour market, all caused by the deep changes spurred by globalisation and consecutive upheavals resulting from ongoing industrial revolutions.

The Recovery Plan for Europe is designed to support three specific fields:

- research and innovation, channelled by Horizon Europe;
- climate and the digital transition, via the Just Transition Fund and Digital Europe Programme;
- preparedness and recovery resilience, through the Recovery and Resilience Facility and new health programme, EU Health.

On the top of these three fields, the Recovery Plan will focus on:

- traditional funding areas such as cohesion policy and the common agriculture policy (CAP), all aimed at guaranteeing stability and modernisation;
- fighting climate change with at least 30% of the EU funds—the highest share in the history of the Union's general budget;
- biodiversity protection and gender equality.

The EU budget is an interventionist tool for the member states. This role stems from a number of solutions that accompany the construction of the budget, the methods applied to collect revenues, and addressing the problem of priorities that can obtain financing. A similar opinion can be formulated in the context of the method used to finance projects approved for financing, as this process requires additional sources from national budgets.

Intervention often awakens the problem of competition because it can create barriers for competitors, which raises the problem of fairness in global competition (level playing field, LPF). The EU factors in a number of conditions, but the most important are trade policies that secure the LPF for companies functioning on the EU market. This can be done through reciprocity in market opening, elimination of unfair practices, keeping high standards, and enforcing the EU's institutional and legal regulations. Conditions such as these help the EU to take advantage of the current and next industrial revolutions (4.0 and 5.0), both driven by digitalisation.

Table 3 Annual trade growth in main sectors with EU preferential partners in 2018–2019 (%)

Contents	All sectors	Agrifood	Industrial	Mineral products	Chemicals	Machinery and appliances	Transport equipment
Imports	2.6	8.3	2.1	−7.1	15.7	3.1	8.4
Exports	4.1	8.7	3.7	−3.1	6.3	1.5	5.7
All trade	3.4	8.5	3.0	−5.8	10.0	2.1	6.8

Source European Commission (2019) European Annual report on implementation of EU trade agreements in 2019, EC, https://ec.europa.eu/commission/presscorner/detail/en/IP_19_6074 (12.11.2020)

The presented data indicate that trade relations within a given group of FTA partners guarantees predictable conditions of trade, which results in lower drops in values of trade in unfavourable conditions like those resulting from the pandemic. Moreover, FTAs create additional advantages in comparison to other trade partners, which can be seen in institutional competitiveness. FTAs and Market Access Partnership both create conditions for sustainable trade relations. This should be interpreted as background for increased reliability, which, while near certain there will be a fall in mutual exchange in a period of economic slowdown, the scale of such a fall is limited in comparison to partners who are not part of such agreements. The European Commission works on the elimination of trade barriers and improvements in the business environment for EU companies engaged in international trade. The observed tendency in trade declines can also be seen in specific groups of traded goods. In general, the data presented below indirectly indicate structural changes in the EU market in consecutive steps towards replacing industrial zombies with competitive and stable sectors (Labor Market, 2020; Perez & McGavin, 2020; Breugel, 2009) (Table 3).

The EU's trade agreements, preceded by negotiations over trade barriers, eliminate or lower tariffs, ease administration procedures, and increases coherency in standards and rules for goods and services. A detailed analysis of the change in trade indicators after signing new agreements shows that mutual turnover takes off, for example, the deal with Japan in which EU goods exports increased by 6% (10% in textiles, clothing, footwear; 13.8% in agri-food). Data concerning trade with Canada (CETA), Ukraine (Free Trade Area), Columbia, Equator and Peru (Association Agreements) also seem to show a similar trend (Annual report on implementation, 2019). The importance of removing trade barriers increases with the structural change of economies, followed by domination of micro, small and medium companies. The trade value of large companies is biggest but their share of all companies functioning on the internal EU market is the smallest. This indicates the importance of reducing obstacles to trade that raise costs and thus limit competitiveness. The size of companies that dominate a given market also indicate changes in the type of produced and exported goods.

Traditionally, as part of the common European external trade policy, the EU enforces its rights through a dispute settlement mechanism (DSM). According to information reported by the Commission, the EU is pursuing 21 complaints within the WTO, which are addressed to 10 different trade partners. The EU and the U.S. are the two leading markets that use the DSM system most extensively. This mechanism and the procedures conducted as part of them is intended to guarantee fair trade deliveries in comparison to total foreign trade transactions, which can be illustrated by raw material imports from China or supplies of refrigerators and paper to Russia (WTO, 2020a, b, c).

Trade Defence Instruments (TDI) are applied to ensure fair competition and are considered important elements of free trade and an open economy. In the EU, there are discussion about how to increase the effectiveness of TDIs. This has resulted in two proposals, one by the Council and the other by the European Parliament. They are pointed at wholesale modernisation, a new anti-dumping calculation methodology tackling the strength of market distortion, and creating the ability to respond to unfair practices in financing subsidies. The mentioned proposals were designed after public consultations and with the support of the EU Member States.

The significance of the problem can be supported also by the fact that nearly 40% of EU exports are conducted with partners institutionally linked with the single market by FTAs (implemented and concluded). The EU is strongly engaged and pushing forward solutions to reform the WTO system, bringing it closer to the current conditions of trade. This also takes into account the Investment Court System and Multilateral Investment Court. The new solutions should also consider that modern production is split internationally across markets located on different continents but within GVCs. Moreover, the divide between industrial goods and services is blurring, which in practice means that a number of exported services results in goods exports. This means that liberalisation of trade in goods, which was expanded by liberalisation of FDI, currently require further expansion into services. Goods trade, FDI flows, and services liberalisation should be accomplished at the same time. There is also an acute need to continue negotiations concerning the Environmental Goods Agreement (EGA) and the Trade in Services Agreement (TiSA). To list should be added talks about the conditions concerning the mobility of professionals.

The continued rise of digital trade has led the Commission to develop a new dedicated FTA chapter on e-commerce for future negotiations to facilitate electronic contracts and transactions, including enhanced consumer protection. Already proposed to Mexico, the EU is advocating exploring the topic further in the WTO. The Commission will continue to analyse the effect of digitalisation on the European economy and identify how trade policy can best reflect these new developments.

All this shows that the engagement of the EU in liberalising trade is not limited to bilateral agreements but covers a wider, global perspective. Taking into account that the US, now under the Biden administration, will also include this topic in its external engagement as part of its foreign policy practise. This will change the trade-war approach of former President Trump towards a more cooperative, international America. This means that the Biden administration will seek more partners abroad than the previous presidency. More detailed indications will come as the Biden cabinet forms and takes control, becoming clearer in the months and years ahead. Nevertheless, Biden's interest in international relations in his work in the U.S.

Senate, is interpreted in such a way that the expectation is that he will look for partners while conducting international policy. This means that there will be a number of new conditions shaping global policy, embracing dimensions in different fields: (1) political; (2) economic; (3) technological; (4) health; and (5) international, shaped by institutions and law (international organisations).

5 Conclusions

Liberalisation of international economic relations develops cyclically, so after a period of growing liberalisation, a return of protectionism can be expected. Protectionism often is provoked by crises. Overcoming a crisis, however, requires international cooperation, which results in further opening of economies. This was experienced most recently in the 2008 financial crises as well as previous downturns in the world economy when states coordinated their policies within the G7, G20, and with advice from the IMF, WB, OECD, EU, etc. The COVID-19 pandemic also requires coordination of certain activities (Phillips, 2020) such as: (1) cooperation in overcoming infections and ending the pandemic; (2) reducing the financial effects of the slowdown in national economies and overcoming the fiscal consequences of intervention costs, which have resulted in swollen budget deficits; (3) the need to prepare the groundwork for companies to recover; (4) a request to reform some international organisations such as the WTO; (5) greening of the economy and advancing the process of the Industrial Revolution 4.0 (and pending Revolution 5.0), marked by significant digitalisation and AI.

All the mentioned conditions can, on the one hand, be seen as background, and yet, on the other, they have prompted higher protectionism, as happened in the past in conditions of an economic slowdown or downturn. Nevertheless, the current conditions are different in several ways, which can be proved by specific features that characterise each country's economy: if it is internationalised and interdependent, this means strong ties to external markets; or, whether an economy is heavily indebted. Structural changes of economies should be coordinated internationally, considering the medium- and longer-term perspective. The changes should include the greening of the world economy, as well as technologies advancing the Industrial Revolutions 4.0 and 5.0.

This analysis has concentrated on developed economies, but the process of effective changes also requires the inclusion of remaining groups of countries that are less advanced in the stages of development. The conducted analysis demonstrates that institutional ties and legal regulations that create predictable conditions for trade between different countries have stabilising effects on trade, which influences positively growth, wealth, and standards of living. External leverage formed by financial sources, know-how, and institutional frames can play an important role in pushing forward solutions that seem to be necessary and effective. Such a coordinated approach can be seen as a solution that will bring a qualitative change in ties between individual economies and their markets, can stimulate the recovery from the COVID-19 pandemic, as well as from the financial consequences of the slow-down and applied policies to overcome the crisis (lockdowns). It can also be seen as a method stimulating technical changes and adjusting the economies and their production to the new challenges brough by consecutive industrial revolutions.

In the past crises, e.g. the 2008 global financial crisis, developing economies seemed to be winners in some areas (FDI flows) in comparison to the most-developed economies. This finding shows that conditions to overcome the current crisis can engage economies at different levels of development. This can mean more invest-ments in catching up and developing economies and deeper structural changes in the most economically advanced markets. In both cases, one can expect a return to a drive that supports the opening of markets. This can be exemplified by struc-tural changes in economies that bring forward new goals, such as greening of the economy, changes in education systems, changes in engagement in GVCs, capital flows, etc. It also requires a new approach within FTAs, followed by wider reforms of the international organisations that form the institutional and legal background for agreements that liberalise the trade of goods and services, flow of capital, and movement of people.

Economic cooperation between states for a long time was limited to a group of countries representing a similar level of development, but now, as practise shows, it has become beneficial for both the developed and developing to work together. This cooperation no longer determines set conditions to sell industrial products manu-factured in developed economies on the markets of less-developed economies. It stimulates investments in less-developed economies, helping them to catch-up and gives them access to products from the markets of developed economies. Devel-oped economies can gain from this division of labour in several ways. They get products while paying lower prices in comparison with those produced on their own markets. They are paid for the intellectual property and services they produce. In both cases, the engagement of production factors brings more added value and makes countries and their citizens wealthier. Investment agreements in this specific process of restructuring the world economy are helpful in taking decisions that are more directed by effectiveness and less by habit and limited knowledge resulting from repeated activities.

Acknowledgements The author would like to thank the SGH Warsaw School of Economics for supporting research reported in this paper through the grant "SGH research potential 2021".

References

Balmford, B., Anan, J.-A., Hargreaves, J.-S., Altoe, M., & Bateman, J. (2020). *Cross country comparisons of COVID-19: Policy, politics and price of life, regulatory quality and COVID-19: Managing the risks and supporting the recovery*. Springer.

Bruegel. (2009). Labor market. The risk of 9 million zombie jobs in Europe. AllianzSE.17.06.2020; Cohen-Setton, J., European Zombies, Bruegel, 29.11.2009.

Burkov, A. (2019). *The hundred page machine learning book*, Andriy Burkov Independently Published.

BusinessWire. (2019). *Global 3D and 4D technology market (2019)—2016–2026—segment analysis opportunity assessment, competitive intelligence, industry outlook*. AII The Research (Featured Publisher), BusinessWire. https://www.businesswire.com/news/home/201911180 05621/en/Global-3D-4D-Technology-Market-Report-2019

Calzada, I., & Almirall, E. (2020). Data ecosystems for protecting European citizens' digital rights. *Transforming Government: People, Process and Policy, 14*(2), 133–147. https://doi.org/10.1108/TG-03-2020-0047

Ciuriak, D., & Ptashkina, M. (2018). *The digital transformation and the transformation of international trade (January 23, 2018). RTA exchange*. International Centre for Trade and Sustainable Development (ICTSD) and the Inter-American Development Bank (IDB). Available at SSRN: https://ssrn.com/abstract=3107811

COM. (2017). Towards a digital trade strategy, EC Brussels 2.11.2017

Edwards, C. (2013). *The fragmented world: Competing perspectives on trade*. Routledge.

European Commission. (2017). Report from the Commission to the European Parliament, The Council, The European Economic and Social Committee and The Committee of Regions. Report on the Implementation of the Trade Policy Strategy Trade for All Delivering a Progressive Trade Policy to Harness Globalisation, European Commission, Brussels, 13.9.2017 COM (2017) 491 final.

European Commission. (2019). European annual report on implementation of EU trade agreements in 2019, EC, (12.11.2020), https://ec.europa.eu/commission/presscorner/detail/en/IP_19_6074

European Commission. (2020a). *Europe investing in digital: The digital European programme with digital innovation hubs and connecting Europe facility (CEF2) digital*. European Commission 2020, (12.11.2020). https://ec.europa.eu/digital-single-market/en/connecting-europe-fac ility-cef2-digital

European Commission. (2020b). Recovery plan for Europe|European Commission (europa.eu) (14.11.2020).

European Union. (2020a). Connecting Europe Facility (CEF2) digital. (12.11.2020). https://ec.eur opa.eu/digital-single-market/en/connecting-europe-facility-cef2-digital

European Union. (2020b). EU trade agreements: Delivering for Europe's businesses, November 2020, (14.11.2020). https://ec.europa.eu/commission/presscorner/detail/en/ip_20_2091

Gandolfo, G. (2014). *International trade theory and policy*. Springer-Verlag.

Goldstein, S. (2020). There's a growing wave of "Zombie" companies in Europe. The EU recovery fund could be a solution. Bank of America Says. The Wall Street Journal, 27 November 2020.

Goodfellow, I. (2016). *Deep learning (adaptive computation and machine learning series)*. MIT Press.

Greenway, D., Falvey, D., Kreickemeier, U., & Berhofen, D. (Eds.). (2013). *Palgrave handbook of international trade*. Palgrave McMillan.

Labor Market. (2020). *World Employment and Social Outlook Trends*. ILO Geneva 2020.

Lytras, M. D., & Visvizi, A. (2021). Artificial intelligence and cognitive computing: Methods, technologies, systems, applications and policy making. *Sustainability, 13*, 3598. https://doi.org/10.3390/su13073598

Pauletto, C. (2020). Information and telecommunications diplomacy in the context of international security at the United Nations. *Transforming Government: People, Process and Policy, 14*(3), 351–380. https://doi.org/10.1108/TG-01-2020-0007

Perez, G. I., McGavin, R. (2020). Europe's "Zombie" borrowers besieged by spread of coronavirus, (6.3.2020).

Phillips, B. (2020). *We know what policies can fix COVID-19 inequality emergency. But only people power can win them. COVID-19, new social contract policy*. OECD, (19.11.2020).

Polese, F., Grimaldi, M., Troisi, O. (2021). Ecosystems transformation for social change: How to challenge emergency through emergence. In C. Leitner, W. Ganz, D. Satterfield & C. Bassano

(Eds) *Advances in the human side of service engineering. AHFE 2021.* Lecture Notes in Networks and Systems, vol. 266. Springer. https://doi.org/10.1007/978-3-030-80840-2_8

Recovery Plan. (2020). European Commission publishes proposal for recovery plan and adjusts 2020 Work Programme, EC, Recovery Plan for Europe, Recovery plan for Europe|European Commission (europa.eu).

Russer, S., & Norvig, P. (2020). *Artificial intelligence: A modern approach.* Prentice Hall.

Sokołowski, M. M. (2021). Regulation in the COVID-19 pandemic and post-pandemic times: Day-watchman tackling the novel coronavirus. *Transforming Government: People, Process and Policy, 15*(2), 206–218. https://doi.org/10.1108/TG-07-2020-0142

Visvizi, A. (2021). Artificial intelligence (AI): Explaining, querying, demystifying. In A. Visvizi & M. Bodziany (Eds.), *Artificial intelligence and its context—Security, business and governance.* Springer.

Visvizi, A., Lytras, M. D. (2019). Politics & ICT: Mechanisms, dynamics, implications. In A. Visvizi & M. D. Lytras (Eds.), *Politics and technology in the post-truth era.* Emerald Publishing.

Visvizi, A., & Lytras, M. (2020). Editorial: Covid-19 transforms the government: Here's what to look at. *Transforming Government: People, Process and Policy, 14*(2), 125–131. https://doi.org/10.1108/TG-05-2020-128

Visvizi, A., Lytras, M. D., & Aljohani, N. (2021). Big data research for politics: Human centric big data research for policy making, politics, governance and democracy. *Journal of Ambient Intelligence and Humanized Computing, 12*(4), 4303–4304. https://doi.org/10.1007/s12652-021-03171-3

WTO. (2018). *World Trade Report 2018: The future of world trade. How digital technologies are transforming global commerce* (pp. 24–59). WTO.

WTO. (2020a). Principles of the trade system. WTO|Understanding the WTO—principles of the trading system, (10.11.2020). www.wto.org/english/thewto_e/whatis_e/tif_e/fact2_e.htm

WTO. (2020b). Trade shows signs of rebound from COVID-19, recovery still uncertain, Press/862 Press Release, (15.11.2020). https://www.wto.org/english/news_e/pres20_e/pr862_e.htm

WTO. (2020c). *World Trade Report: Government policies to promote innovation in the digital age* (pp. 22–74). World Trade Organization (WTO). https://www.wto.org/english/res_e/publications_e/wtr20_e.htm

Żukrowska, K. (Ed.). (2013). *Kryzys gospodarczy 2008+ Test dla stosowanej polityki [Economic crisis 2008+. Testing the economic policy measures applied].* Oficyna Wydawnicza SGH

Katarzyna Żukrowska, Professor of economics and political sciences with the specialisation of international economic and political relations, Institute of International Studies, College of Socio-Economics, SGH Warsaw School of Economics. ORCID: 000-0001-6751-5760; Email: kzukro@sgh.waw.pl

Printed in the United States
by Baker & Taylor Publisher Services